ASCE/SEI 43-05

American Society of Civil Engineers

Seismic Design Criteria for Structures, Systems, and Components in Nuclear Facilities

This document uses both the International System of Units (SI) and customary units.

ASCE

SEI
Structural Engineering Institute
of the American Society of Civil Engineers

Published by the American Society of Civil Engineers

Library of Congress Cataloging-in-Publication Data

Structural Engineering Institute. Working Group for Seismic Design Criteria for Nuclear Facilities.

Seismic design criteria for structures, systems, and components in nuclear facilities developed by Working Group for Seismic Design Criteria for Nuclear Facilities, Dynamic Analysis of Nuclear Structures Subcommittee, Nuclear Standards Committee.

p. cm.

"This document uses both the International System of Units (SI) and customary units."

Includes bibliographical references and index.

ISBN 0-7844-0762-2

1. Nuclear power plants—Earthquake effects. 2. Earthquake resistant design. I. Structural Engineering Institute. Dynamic Analysis of Nuclear Structures Subcommittee. II. Structural Engineering Institute. Nuclear Standards Committee. III. Title.

TK9152.163.S77 2005
621.48′32—dc22

2005005011

Published by American Society of Civil Engineers
1801 Alexander Bell Drive
Reston, Virginia 20191
www.pubs.asce.org

Library of Congress Catalog Card No.: 2005005011
ISBN 0-7844-0762-2
Manufactured in the United States of America.

STANDARDS

In April 1980, the Board of Direction approved ASCE Rules for Standards Committees to govern the writing and maintenance of standards developed by the Society. All such standards are developed by a consensus standards process managed by the Codes and Standards Activities Committee. The consensus process includes balloting by the Balanced Standards Committee, which is composed of Society members and nonmembers, balloting by the membership of ASCE as a whole, and balloting by the public. All standards are updated or reaffirmed by the same process at intervals not exceeding 5 years.

The following Standards have been issued:

- ANSI/ASCE 1-82 N-725 Guideline for Design and Analysis of Nuclear Safety Related Earth Structures
- ANSI/ASCE 2-91 Measurement of Oxygen Transfer in Clean Water
- ANSI/ASCE 3-91 Standard for the Structural Design of Composite Slabs and ANSI/ASCE 9-91 Standard Practice for the Construction and Inspection of Composite Slabs
- ASCE 4-98 Seismic Analysis of Safety-Related Nuclear Structures
- Building Code Requirements for Masonry Structures (ACI 530-02/ASCE 5-02/TMS 402-02) and Specifications for Masonry Structures (ACI 530.1-02/ASCE 6-02/TMS 602-02)
- SEI/ASCE 7-02 Minimum Design Loads for Buildings and Other Structures
- ANSI/ASCE 8-90 Standard Specification for the Design of Cold-Formed Stainless Steel Structural Members
- ANSI/ASCE 9-91 listed with ASCE 3-91
- ASCE 10-97 Design of Latticed Steel Transmission Structures
- SEI/ASCE 11-99 Guideline for Structural Condition Assessment of Existing Buildings
- ANSI/ASCE 12-91 Guideline for the Design of Urban Subsurface Drainage
- ASCE 13-93 Standard Guidelines for Installation of Urban Subsurface Drainage
- ASCE 14-93 Standard Guidelines for Operation and Maintenance of Urban Subsurface Drainage
- ASCE 15-98 Standard Practice for Direct Design of Buried Precast Concrete Pipe Using Standard Installations (SIDD)
- ASCE 16-95 Standard for Load and Resistance Factor Design (LRFD) of Engineered Wood Construction
- ASCE 17-96 Air-Supported Structures
- ASCE 18-96 Standard Guidelines for In-Process Oxygen Transfer Testing
- ASCE 19-96 Structural Applications of Steel Cables for Buildings
- ASCE 20-96 Standard Guidelines for the Design and Installation of Pile Foundations
- ASCE 21-96 Automated People Mover Standards—Part 1
- ASCE 21-98 Automated People Mover Standards—Part 2
- ASCE 21-00 Automated People Mover Standards—Part 3
- SEI/ASCE 23-97 Specification for Structural Steel Beams with Web Openings
- SEI/ASCE 24-98 Flood Resistant Design and Construction
- ASCE 25-97 Earthquake-Actuated Automatic Gas Shut-Off Devices
- ASCE 26-97 Standard Practice for Design of Buried Precast Concrete Box Sections
- ASCE 27-00 Standard Practice for Direct Design of Precast Concrete Pipe for Jacking in Trenchless Construction
- ASCE 28-00 Standard Practice for Direct Design of Precast Concrete Box Sections for Jacking in Trenchless Construction
- SEI/ASCE/SFPE 29-99 Standard Calculation Methods for Structural Fire Protection
- SEI/ASCE 30-00 Guideline for Condition Assessment of the Building Envelope
- SEI/ASCE 31-03 Seismic Evaluation of Existing Buildings
- SEI/ASCE 32-01 Design and Construction of Frost-Protected Shallow Foundations
- EWRI/ASCE 33-01 Comprehensive Transboundary International Water Quality Management Agreement
- EWRI/ASCE 34-01 Standard Guidelines for Artificial Recharge of Ground Water
- EWRI/ASCE 35-01 Guidelines for Quality Assurance of Installed Fine-Pore Aeration Equipment
- CI/ASCE 36-01 Standard Construction Guidelines for Microtunneling
- SEI/ASCE 37-02 Design Loads on Structures During Construction
- CI/ASCE 38-02 Standard Guideline for the Collection and Depiction of Existing Subsurface Utility Data
- EWRI/ASCE 39-03 Standard Practice for the Design and Operation of Hail Suppression Projects
- ASCE/EWRI 40-03 Regulated Riparian Model Water Code
- ASCE/EWRI 42-04 Standard Practice for the Design and Operation of Precipitation Enhancement Projects
- ASCE/SEI 43-05 Seismic Design Criteria for Structures, Systems, and Components in Nuclear Facilities
- ASCE/EWRI 44-05 Standard Practice for the Design and Operation of Supercooled Fog Dispersal Projects

CONTENTS

Commentary

resistance and the associated deformation shall include [illegible] friction and cohesion as appropriate.

The horizontal deflection of the foundation is determined using the load-versus-deflection curve for the combined effect of cohesion, friction, and passive resistance. The [illegible] of the building is checked, and if it is not impaired, an acceptable condition exists.

If the above condition is not satisfied, it is possible to demonstrate an acceptable condition by estimating the larger displacement resulting when the dynamic base shear momentarily exceeds the soil resistance and demonstrating the building can remain functional for the resulting displacements. A refined estimate of the foundation movement shall recognize the time-dependent reversing nature of earthquake motion and the nonlinear soil resistance. If the functionality of the building can be demonstrated for the larger displacements, an acceptable condition exists.

The nonlinear sliding analysis method of Section [illegible] of Appendix A of this Standard can be used to estimate the foundation sliding movement as long as the fundamental frequency [illegible] of the building exceeds the effective cutoff frequency [illegible] defined in Section [illegible]. When [illegible] the appropriate coefficient of friction [illegible] to be used in Eq. A-1 is given by

[illegible] (Eq. [illegible])

where

V = Total sliding resistance
W = Building weight

[illegible] Building Overturning

The [illegible] criteria for overturning is [illegible] when overturning moment [illegible] analysis is [illegible] the overturning moment, a [illegible] provides a reasonable estimate of the [illegible] pressure. Therefore, building stability [illegible].

[illegible]

FIGURE [illegible] Rigid Body Rocking Definitions

[illegible] analysis is used, a minimum of [illegible] that satisfy the requirements [illegible] shall be used. The mean values of sliding [illegible] from these multiple [illegible] shall be used as the "best-estimate" values.

[illegible] BUILDING SLIDING AND OVERTURNING

[illegible] Building Sliding

[illegible] sliding stability of a building can be checked [illegible] comparing the lateral force [illegible] resistance.

[illegible]

[illegible] demonstrated by satisfying the [illegible] from linear analysis is determined [illegible] given in Section [illegible] for the [illegible] the foundation, the static resistance [illegible]

[illegible]

FOREWORD

Nuclear facilities are defined as facilities that process, store, or handle radioactive materials in a form and quantity that pose potential nuclear hazard to the workers, the public, or the environment. Due to the risk associated with such hazards, it is desirable that nuclear facilities have a lower probability that structural damage will be caused by earthquakes than do conventional facilities. This Standard provides seismic design criteria that are more stringent than normal building codes. The goal of this Standard is to ensure that nuclear facilities can withstand the effects of earthquake ground shaking with desired performance, expressed as probabilistic Target Performance Goals. Design for other earthquake effects (such as differential fault displacement and seismic slope instability) are not covered by this Standard. This Standard is intended for use in the design of new facilities and is to be used in conjunction with other national consensus standards specified herein.

This Standard can also be used for facilities handling explosives, toxic materials, or chemicals; for facilities where safety, mission, or investment protection are concerns; and where more stringent seismic criteria than provided by building codes are desired.

This Standard is intended to be used with ASCE 4, which provides criteria for seismic analysis of safety related nuclear facilities Structures, Systems and Components (SSCs); ACI 349 for concrete structures; AISC standards for steel structures; ASME standards for mechanical systems and components; IEEE standards for electrical systems and components; and ASCE 7 for minimum non-seismic design loads for buildings and other structures. This ASCE Standard specifies seismic load combinations.

This Standard uses the Target Performance Goal–based seismic design approach documented in U.S. Department of Energy Natural Phenomena Hazards (NPH) standards. This Standard is also consistent with the philosophy used in the National Earthquake Hazard Reduction Program (NEHRP) for seismic mitigation of new and existing facilities. The Standard uses input from ANSI/ANS Standard 2.26 to assign Seismic Design Categories (SDCs)* to SSCs. It provides requirements for determining design basis seismic loading using input from ANSI/ANS Standards 2.27 and 2.29, and it prescribes design criteria that are tied to structural Limit States.

ANS 2.26 employs a graded approach to ensure that the level of conservatism and rigor in design is appropriate for facility characteristics, such as hazards to workers, the public, and the environment. ANS 2.26 specifies five SDCs for classifying SSCs based on their importance and failure consequences. Each SSC has a specified numerical Target Performance Goal. ANS 2.26 also provides descriptive criteria to assist the designer in selecting an appropriate Limit State for use in the design of SSCs. Four Limit States are defined—A, B, C, and D—where A is short of collapse and D is essentially elastic behavior. This Standard specifies design criteria for load combinations, including earthquake ground shaking (i.e., stress, displacement, and ductility limits), such that these Limit States are not exceeded.

The combination of SDC and Limit State defines the Seismic Design Basis (SDB) for each SSC. Thus, an SSC with SDB-3C would use criteria for SDC-3 and Limit State C. A total of 20 SDBs are defined in ANS 2.26 that can match seismic design criteria to SSC safety function and importance, implementing a graded approach.

SDBs defined by SDC 1 and 2 are covered by the approach presented in ASCE 7. This Standard presents design and analysis requirements for SDBs defined by SDC 3, 4, and 5 and all Limit States. The approach presented for SDC 3, 4, and 5 has been adapted from that used in the U.S. Department of Energy Standard 1020, ASCE 4, and the U.S. Nuclear Regulatory Commission Standard Review Plan (NUREG-0800).

The intended user of this Standard is the designer or analyst involved in the design of a new nuclear structure, system, or component. The Standard is intended to provide a rational basis for the performance-based, risk-consistent seismic design of SSCs in nuclear facilities. Designers once were initiated into the field of probabilistic design by being taught that seismic performance categories for SSCs were established by DOE-STD-1020-94 and subsequent revisions. Each performance category was tied to a probabilistic performance goal that represented a target annual frequency of seismic-induced failure. However, these earlier design codes did not allow designers the freedom to select a Limit State (the permissible deformation limit for the SSC established from functional considerations). There has been a movement within the structural engineering community to give designers freedom to select the desired state of the facility following the Design Basis Earthquake (DBE, defined in ATC-40, FEMA 273 and FEMA 356, SEAOC-Vision 2000, and ASCE 31). The traditional design Limit State of providing life safety can now be expanded to include nuclear confinement, remain fully functional, or minimize operational loss.

* In this Standard, the term "Seismic Design Category" has a different meaning than in the International Building Code and ASCE 7.

ACKNOWLEDGMENTS

The American Society of Civil Engineers acknowledges the work of the Nuclear Standards Committee of the Codes and Standards Activities Division of the Structural Engineering Institute. This group comprises individuals from many backgrounds, including consulting engineering, law, research, construction industry, education, government, design, and private practice.

This Standard was prepared through the consensus standards process by balloting in compliance with procedures of the Codes and Standards Activities Committee of the American Society of Civil Engineers. Those individuals who served on the Nuclear Standards Committee are:

Nuclear Standards Committee

S. Bolourchi
J. Costello
C. Costantino
F. Feng
O. Gurbuz
A. Hadjian
R. Kassawara
R. Kennedy
W. La Pay
R. C. Murray, Chair
T. Satyan-Sharma
P. Wang

Working Group for Seismic Design Criteria for Nuclear Facilities

G. Bagchi
H. Chander
C. J. Costantino
M. D. Davister
R. C. Guenzler
O. Gurbuz
A. H. Hadjian
Q.A. Hossain
T.W. Houston
R. J. Hunt
R.P. Kennedy
L. Manuel
G. E. Mertz
D. P. Moore
R. C. Murray
D. Neihoff
T. A. Nelson, Cochair
M. E. Nitzel
D. A. Nuta, Cochair
J. W. Reed
M. J. Russell
M. W. Salmon
S. K. Sen
S. A. Short
W. H. White

Dynamic Analysis of Nuclear Structures Subcommittee

G. Bagchi
H. Chander
C. J. Costantino
M. D. Davister
R. C. Guenzler
O. Gurbuz
A. H. Hadjian
Q.A. Hossain
T.W. Houston
R. J. Hunt
R.P. Kennedy
L. Manuel
G. E. Mertz
D. P. Moore
R. C. Murray, Chair
D. Neihoff
T. Nelson
M. E. Nitzel
D. Nuta
J. W. Reed
M. J. Russell
M. W. Salmon
S. K. Sen
S. A. Short
W. H. White

ACRONYMS/NOTATION

A_I	Arias intensity
A_R	Ground motion ratio
ACI	American Concrete Institute
AISC	American Institute of Steel Construction
AISI	American Iron and Steel Institute
ANS	American Nuclear Society
ANSI	American National Standards Institute
APE	Annual probability of exceedance
ASD	Allowable Stress Design
ASME	American Society of Mechanical Engineers
ATC	Applied Technology Council
AWWA	American Water Works Association
B&PVC	Boiler and Pressure Vessel Code
C	Capacity determined in accordance with building codes
CMAA	Crane Manufacturer Association of America
COV	Coefficient of variation
D	Total demand; also, distance to controlling earthquake; also, peak ground displacement
D_{NS}	Non-seismic demand
D_S	Elastic seismic demand
DBE	Design Basis Earthquake
DF	Design Factor
DOE	U.S. Department of Energy
DRS	Design Earthquake Response Spectrum: $DRS = DF \times UHRS$
EBF	Eccentrically Braced Frame
EES	Earthquake Experience Spectrum
ENA	Eastern North America
EUS	Eastern United States
EPRI	Electric Power Research Institute
LRFD	Load and Resistance Factor Design
F_μ	Inelastic energy absorption factor
$F_{\mu S}$	System inelastic energy absorption factor
FEMA	Federal Emergency Management Agency
FS	Factor of Safety
GIP	Generic Implementation Procedure
H_D	Mean annual hazard exceedance frequency: $H_D = R_P \times P_F$
IBC	International Building Code
IEEE	The Institute of Electrical and Electronics Engineers, Inc.
K	Capacity increase factor
LS	Limit State (A, B, C, or D)
M	Magnitude of controlling earthquake
N_y	Nyquist frequency
NEHRP	National Earthquake Hazard Reduction Program
NEMA	National Electrical Manufacturer Association
NEP	Non-Exceedance Probability
NFPA	National Fire Protection Association
NPH	Natural Phenomena Hazards
NPP	Nuclear Power Plant
NRC	U.S. Nuclear Regulatory Commission
PC	Performance Category
P_F	Mean annual frequency of unacceptable performance (Target Performance Goal)
PGA	Peak Ground Acceleration; A is also used for Peak Ground Acceleration
PSD	Power Spectral Density
PSHA	Probabilistic Seismic Hazard Assessment
QA	Quality Assurance
R_P	Probability Ratio: H_D / P_F
RBS	Reduced Beam Sections
RRS	Required Response Spectra
SA_f	Spectral Acceleration at natural frequency, f
SA_{PEAK}	Peak Spectral Acceleration
SAM	Seismic Anchor Motion
SDB	Seismic Design Basis
SDC	Seismic Design Category* (SDC-1, SDC-2, SDC-3, SDC-4, or SDC-5)
SF	Seismic Scale Factor
SMACNA	Sheet Metal and Air-Conditioning Contractors National Association
SMRF	Special Moment-Resisting Frame
SQUG	Seismic Qualification Utility Group
SRSS	Square root sum of squares
SSC	Structure, System, or Component
SSE	Safe Shutdown Earthquake
SSI	Soil-Structure Interaction
T_{sm}	Strong motion duration
TES	Test Experience Spectrum
TRS	Test Response Spectrum
UHRS	Uniform Hazard Response Spectra
USGS	U.S. Geological Survey
V	Peak Ground Velocity
ZPA	Zero Period Acceleration
α	Parameter used to determine Design Factor
ϕ	Capacity reduction factor

DEFINITIONS

ACCELEROGRAM. A representation (either recorded, modified recorded, or synthetic) of the acceleration of the ground during an earthquake. The accelerogram contains acceleration and time-data pairs.

RECORDED EARTHQUAKE ACCELEROGRAM. A time-history record of acceleration versus time that has been measured by a strong motion instrument during an earthquake.

MODIFIED RECORDED EARTHQUAKE ACCELEROGRAM. A time-history record of acceleration versus time that has been produced from a recorded earthquake time history, but in which the Fourier amplitudes have been scaled such that the resulting response spectrum envelops a target response spectrum. The Fourier phasing from the Recorded Earthquake Accelerogram is preserved in a Modified Recorded Earthquake Accelerogram.

SYNTHETIC EARTHQUAKE ACCELEROGRAM. A time-history record of acceleration versus time pairs that has been produced so that the resulting response spectrum envelops a target response spectrum.

ACTIVE COMPONENT. Components that must change state as part of their safety function during an earthquake.

ARIAS INTENSITY. A measure of the intensity of ground shaking that is obtained by integrating the square of the ground acceleration values over a specified time period. The Arias intensity is given as

$$A_I = \frac{\pi}{2g} \cdot \int_0^{t_m} a^2(t)dt$$

where $a(t)$ is the ground acceleration and t_m is the duration of the ground acceleration record.

ARIAS INTENSITY RISE TIME. Duration (time) needed to produce 5% of the total cumulative energy available in an earthquake accelerogram. If the total cumulative energy, E_{total}, is given as

$$E_{\text{total}} = \int_0^{\infty} a^2(t)dt$$

then the Arias Intensity Rise Time, $T_{0.05}$, is given as

$$(0.05)E_{\text{total}} = \int_0^{T_{0.05}} a^2(t)dt$$

BACKBONE CURVE. Monotonic representation of the nonlinear response of an element under consideration obtained by enveloping the load deformation curve of the element.

CAPACITY SPECTRUM METHOD. A nonlinear static analysis procedure (described in ATC-40) that provides a graphical representation of the expected seismic performance of a structure by the intersection of the structure's capacity spectrum with a response spectrum (demand spectrum) representation of the earthquake's displacement demand on the structure. The intersection is the performance point, and the displacement coordinate, d_p, of the performance point is the estimated displacement demand on the structure for the specified level of seismic hazard.

CONTROLLING EARTHQUAKE. The earthquake, for the particular return period and structural frequency range of interest, generated from the deaggregated hazard analysis for the predominant magnitude, M, and distance, D, pair.

DESIGN BASIS EARTHQUAKE (DBE). The description of the ground motion, defined in terms of the DRS, to be used for design. The DBE is obtained by following ANSI/ANS 2.27 and 2.29 for SSCs in SDC 3, 4, or 5.

DESIGN RESPONSE SPECTRA (DRS). Response spectra used for design. The DRS are equal to the product of the UHRS and the Design Factor and are defined at a control location in the free field.

DESIGN FACTOR (DF). The ratio between the DRS and the UHRS. The Design Factor is aimed at achieving the target annual probability of failure goals.

DESIGN TEAM. The responsible group charged with producing the design. Typically consists of a number of discipline-specific team members (e.g., structural, mechanical, electrical).

DIRECTIONAL CORRELATION COEFFICIENT. A measure of the degree of linear relationship between two earthquake accelerograms. For accelerograms X and Y, the directional correlation coefficient is given by

$$\rho_{XY} = \frac{\frac{1}{n}\sum_{i=1}^{n}[(X_i - \bar{x})(Y_i - \bar{y})]}{\sigma_X \sigma_Y}$$

where n is the number of discrete acceleration-time data points, $\bar{x}$ and $\bar{y}$ are the mean values, and σ_X and σ_Y are the standard deviations of X and Y, respectively.

* In this Standard, the term "Seismic Design Category" has a different meaning than in the International Building Code and ASCE 7.

DISTRIBUTION SYSTEMS. A system (i.e., collection of components) whose function is to distribute material/data (fluid, signals, power). Examples are piping, cable trays, conduit, and HVAC systems.

DOMINANT RESPONSE PARAMETER. The mode of behavior of the structural component that has the largest contribution to deflection. For example, shear is the dominant response parameter for a squat shear wall [aspect ratio (height/length) less than 2].

EFFECTIVE NATURAL FREQUENCY. The frequency of the single mode of response that dominates the structure or component response for multi–degree-of-freedom structures.

EFFECTIVE STIFFNESS FACTOR. Modifier (e.g., 0.5, 0.7) that is applied to the uncracked section properties of a reinforced concrete member to account for the softening effect that cracking has on the uncracked stiffness properties of interest.

EFFECTIVE STRUCTURAL FREQUENCY. See *Effective Natural Frequency.*

FACILITY. One or more buildings or structures, including systems and components, dedicated to a common function.

FOUNDATION ELEMENT. A structural component that is dedicated to transferring loads from the superstructure to the supporting soil.

FOURIER AMPLITUDE SPECTRUM. A plot of Fourier amplitude, $F(\omega)$, versus frequency, ω. $F(\omega)$ is the Fourier amplitude of the time history computed over the strong motion duration, T_{sm}.

SMOOTHED FOURIER AMPLITUDE SPECTRUM. An averaged Fourier amplitude spectrum, computed by averaging the amplitude values $F(\omega)$ over the frequency range of $\omega_i \pm 20\%$ at each frequency point, ω_i, over a moving frequency window.

FOURIER PHASE SPECTRUM. A plot of Fourier phase ϕ versus frequency, ω.

GRADED APPROACH. (From CFR 830.3.) The process of ensuring that the level of analysis, documentation, and actions used to comply with a requirement are commensurate with the following:

- Relative importance to safety, safeguards, and security;
- Magnitude of any hazard involved;
- Life cycle stage of a facility;
- Programmatic mission of a facility;
- Particular characteristics of a facility;
- Relative importance of radiological and nonradiological hazards; and
- Any other relevant factor.

GROUND MOTION SLOPE RATIO. Ratio of the spectral accelerations, frequency by frequency, from a seismic hazard curve corresponding to a 10-fold reduction in hazard exceedance frequency (see Eq. 2.2-2).

HAZARD. A source of danger (i.e., material, energy source, or operation) with the potential to cause illness, injury, or death to personnel (workers or the public), damage to an operation, or damage to the environment (without regard for the likelihood or credibility of accident scenarios or consequence mitigation).

HAZARD CURVE. See *Seismic Hazard Curve.*

HYSTERESIS LOOP. Nonlinear load-deformation loop of a structural component. The area enclosed by hysteresis loop is equivalent to the energy dissipated by the element in one complete loading and unloading cycle.

INELASTIC ENERGY ABSORPTION FACTOR (F_μ). A reduction factor used to reduce demand to account for inelastic behavior. The inelastic energy absorption factor is a function of the Limit State and the structural system or equipment configuration (see Tables 5-1 and 8-1).

IN-STRUCTURE RESPONSE SPECTRA. The response spectra generated from the seismic response at selected locations in a structure. In-structure response spectra are used for design of systems and components supported within a structure.

LIMIT STATE (LS). The limiting acceptable condition of the SSC. The Limit State may be defined in terms of a maximum acceptable displacement, strain, ductility, or stress. Four Limit States are specified in this Standard:

A = Short of collapse, but structurally stable
B = Moderate permanent deformation
C = Limited permanent deformation
D = Essentially elastic

LOAD PATH. The path of resistance consisting of structural or nonstructural members that the imposed load will follow from the point of origin (inertial forces at location of structure mass) to the point of final resistance (e.g., supporting soil).

MEAN ANNUAL HAZARD EXCEEDANCE FREQUENCY. The expected annual probability of exceedance. This value is used to determine earthquake acceleration from seismic hazard curves.

MEAN ANNUAL EXCEEDANCE FREQUENCY OF ACCEPTABLE PERFORMANCE. See *Target Performance Goal.*

NUCLEAR FACILITY. Includes both reactor and nonreactor facilities.

NONREACTOR NUCLEAR FACILITY. Facilities that contain activities or operations that involve radioactive and/or fissionable materials in such form and quantity that a nuclear hazard potentially exists to the employees, the general public, or the environment. Included are activities or operations that:

- Produce, process, or store radioactive liquid or solid waste, fissionable materials, or tritium;
- Conduct separations operations;
- Conduct irradiated materials inspection, fuel fabrication, decontamination, or recovery operations;
- Conduct fuel enrichment operations;
- Perform environmental remediation or waste management activities involving radioactive materials.

Linear accelerators and targets are considered nonreactor nuclear facilities. Incidental use and generation of radioactive materials in a facility operation (e.g., check and calibration sources and use of radioactive sources in research, experimental, and analytical laboratory activities, electron microscopes, and X-ray machines) would not ordinarily require the facility to be included in this definition.

NYQUIST FREQUENCY (N_y). Maximum frequency that can be represented by the time history discretization, $N_y = 1/(2\Delta t)$, where Δt is the time increment.

OVERDRIVE. Using too strong of a time history as input to the soil column so that nonlinear effects result in a larger soil response (soil strains) than would occur under the appropriate event. Overdriving the soil column could produce significantly different response spectra in the free field.

***P*-DELTA (*P*-Δ) EFFECT.** Additional moment induced in axial load-carrying members caused by lateral structural deformation. The *P*-Delta moment is the product of the axial force and the relative lateral displacement between the end points of the member.

PASSIVE COMPONENT. Components that do not require changing state as part of their safety function during an earthquake.

PEAK GROUND ACCELERATION (PGA). The maximum absolute value of the ground acceleration time history.

PEAK SPECTRAL ACCELERATION The maximum acceleration response that a prescribed forcing function can produce in a single-degree-of-freedom oscillator (independent of the natural frequency of the oscillator).

PEER REVIEW. A formal review process in which an external party will review the methodology, results, and process by which a design is developed. The external party is independent of project schedule and budget constraints.

PERFORMANCE GOAL. The mean annual frequency of unacceptable performance that is specified, as a target, for the SSC SDC. Performance goals are specified in ANS 2.26.

PHASE SPECTRUM. See *Fourier Phase Spectrum.*

PINCHED HYSTERETIC BEHAVIOR. A characteristic of the load-deformation loop of a structural component subjected to cyclic loading that is marked by both strength and stiffness degradation in successive loading and unloading cycles beyond yield. See example below.

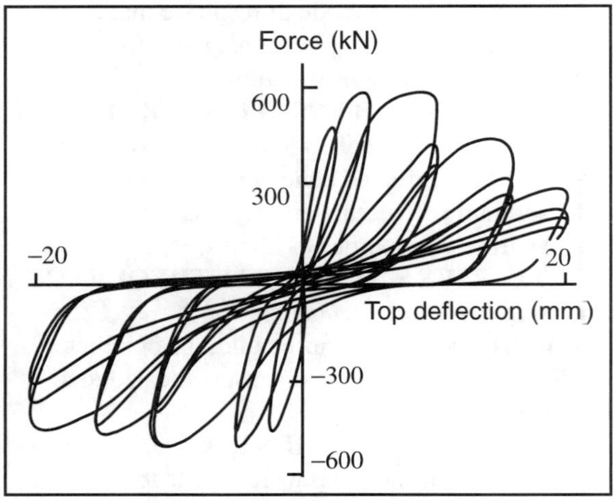

PLASTIC HINGE LENGTH. Region of plastic deformation; may be approximated by 1 beam depth.

POWER SPECTRAL DENSITY (PSD). A measure of the distribution of power in an accelerogram as a function of frequency. The PSD computed from an accelerogram is defined in terms of the Fourier amplitudes of the time history, $F(\omega)$, by the relation

$$PSD(\omega) = \frac{2|F(\omega)|^2}{2\pi T_{sm}}$$

where T_{sm} is the strong motion duration.

PROBABILITY RATIO (R_P). The ratio between the exceedance frequency of the DBE and the Target Performance Goal. The user may specify a unique value for a specific application.

PROBABILISTIC SEISMIC HAZARD ASSESSMENT (PSHA). A procedure used to develop seismic hazard curves and uniform hazard response spectra for determining the ground motion at a site to be used for seismic design. Criteria and guidance for conducting a PSHA are provided in ANSI/ANS 2.27 and 2.29.

REQUIRED RESPONSE SPECTRA (RRS). The representation of the response spectra that are required to qualify an SSC. The required response spectra will include factors required to meet probabilistic performance goals.

SAFE SHUTDOWN EARTHQUAKE (SSE). DBE for commercial nuclear power plants. Per 10 CFR 100, Appendix A, the SSE is that earthquake which is based upon an evaluation of the maximum earthquake potential considering the regional and local geology and seismology and specific characteristics of local subsurface material. It is that earthquake which produces the maximum vibratory ground motion for which certain SSCs are designed to remain functional. These SSCs are those necessary to ensure the following:

- Integrity of the reactor coolant pressure boundary,
- Capability to shut down the reactor and maintain it in a safe shutdown condition, or
- Capability to prevent or mitigate the consequences of accidents, which could result in potential offsite exposures.

The SSE is developed by the nuclear power plant owner and reviewed by the U.S. Nuclear Regulatory Commission.

SEISMIC DEMAND. The demand imposed on the SSC being evaluated at the earthquake level under consideration. The seismic demand may be expressed in terms of force, moment, stress, displacement, rotation, or strain.

SEISMIC DESIGN BASIS (SDB). The combination of SDC (3, 4, or 5) and Limit State (A, B, C, or D) that determines the DBE and acceptance criteria for designing SSCs. For example, SDB-3C would use criteria given in this Standard for SDC-3 and Limit State C.

SEISMIC DESIGN CATEGORY (SDC). A category assigned to an SSC that is a function of the severity of adverse radiological and toxicological effects of the hazards that may result from the seismic failure of the SSC on workers, the public, and the environment. SSCs may be assigned to SDCs that range from 1 to 5. For example, a conventional building whose failure may not result in any radiological or toxicological consequences is assigned to SDC-1; a safety-related SSC in a nuclear material processing facility with a large inventory of radioactive material may be placed in SDC-5. In this Standard, the term "Seismic Design Category" has a different meaning than in the International Building Code and Standard 7. The definition from ASCE 7 follows: "A classification assigned to a structure based on its Seismic Use Group and the severity of the design earthquake ground motion at the site." ANS 2.26 provides guidance on the assignment of SSCs to SDCs.

SEISMIC HAZARD CURVE. Description of the ground motion parameter of interest as a function of annual frequency of exceedance. The seismic hazard curve is determined from a probabilistic seismic hazard assessment following the guidance in ANSI/ANS 2.27 and 2.29.

SLAB/WALL MOMENT FRAME. A moment-resisting frame, composed of both walls and slabs, that resists seismic lateral loading by out-of-plane bending. Slab/wall moment frames may include column and beam elements. The span of the slab is predominately one-way from wall to wall, although two-way action is utilized for concentrated loads and around floor openings. Out-of-plane bending of the walls and slabs resists both gravity and lateral loads. Longitudinal loads are resisted by in-plane shear in the slabs and shear walls. Reinforced concrete structures, which resist lateral seismic load, in two orthogonal directions, with shear walls and diaphragms, are not slab/wall moment frames.

SPECTRAL ACCELERATION (SA). The maximum acceleration response of a single-degree-of-freedom oscillator with a known frequency, f, and viscous damping, β, subjected to a prescribed forcing function or earthquake ground motion time history.

SPECIAL MOMENT-RESISTING FRAME (SMRF). A steel or reinforced concrete moment-resisting frame specially detailed to provide ductile behavior that complies with the requirements given in ANSI/AISC 341-02, or with the special seismic provisions of ACI 349.

STRONG MOTION DURATION (T_{sm}). The duration (in seconds) in which the cumulative energy in an accelerogram moves from 5% to 75% of the total cumulative energy. See *Arias Intensity Rise Time*.

STRUCTURE, SYSTEM, OR COMPONENT (SSC). A *structure* is an element, or a collection of elements, to provide support or enclosure, such as a building, free-standing tanks, basins, dikes, or stacks. A *system* is a collection of components assembled to perform a function, such as piping, cable trays, conduits, or HVAC. A *component* is an item of mechanical or electrical equipment, such as a pump, valve, or relay, or an element of a larger array, such as a length of pipe, elbow, or reducer. In this Standard, each SSC is assigned an SDB that is based on the SDC and the Limit State that are determined following the guidance contained in ANSI/ANS 2.26.

TARGET DISPLACEMENT METHOD. A nonlinear static analysis procedure (described in FEMA 356) that provides a numerical process for estimating the displacement demand on the structure. A bilinear representation of the capacity curve and a series of modification factors, or coefficients, are used to calculate a target displacement. The point on the capacity curve at the target displacement is the equivalent of the performance point in the capacity spectrum method. See *Capacity Spectrum Method*.

TARGET PERFORMANCE GOAL (P_F). Target annual frequency of exceeding a specified Limit

State. Performance goals of 1×10^{-4}, 4×10^{-5}, and 1×10^{-5} are established in this Standard for SDC-3, SDC-4, and SDC-5, respectively. For example, the expected probability of exceeding a Limit State in SDC-3 in any given year is less than 1/10,000. The user may specify, with justification, a unique value for a specific application.

UNIFORM HAZARD RESPONSE SPECTRA (UHRS). Response spectra derived so that the annual probability of exceeding the spectral quantity (acceleration, displacement, etc.) is the same for any spectral frequency. Determined in accordance with ANSI/ANS 2.27 and 2.29.

ZERO PACKING. The practice of lengthening the total duration of an earthquake accelerogram by adding values of zero acceleration to the beginning or the end of the record for the purpose of performing discrete Fourier analysis.

ZERO PERIOD ACCELERATION (ZPA). The maximum absolute value of the ground or in-structure acceleration time-history record.

Seismic Design Criteria for Structures, Systems, and Components in Nuclear Facilities

SECTION 1.0 INTRODUCTION

SECTION 1.1 OVERVIEW OF THE SEISMIC DESIGN CRITERIA

This Standard provides criteria for seismic design of safety-related Structures, Systems, and Components (SSCs) in a broad spectrum of nuclear facilities. Notation, definitions, guidelines, and commentary material are included.

A graded approach is used in developing the seismic design criteria presented in this Standard. The intent is to control the design process such that the performance of the SSC related to safety and environmental protection is acceptable.

To implement the graded approach for seismic design, 20 Seismic Design Bases (SDBs) have been defined as specified in ANS 2.26 and as shown in Table 1-1. SDBs have a quantitative probabilistic Target Performance Goal, P_F, defined for each Seismic Design Category (SDC)* and a qualitative goal defined for each Limit State (LS) or level of acceptable structural behavior. SDBs defined by SDC 1 and 2 are covered by the approach presented in the International Building Code (IBC). In the future, the IBC will use ASCE 7 for its seismic criteria. This Standard, ASCE/SEI 43-05, presents design and analysis requirements for SDBs defined by SDC 3, 4, and 5 and all Limit States.

Target quantitative performance goals decrease in annual probability of exceeding acceptable behavior limits as the SDC increases from 1 to 5. Decreasing quantitative Target Performance Goals are achieved in this Standard by increasing seismic demand associated with the Design Basis Earthquake (DBE). The DBE is defined by Uniform Hazard Response Spectra (UHRS) determined at a specified annual probability of exceedance, H_D, multiplied by a Design Factor (DF). The UHRS annual probability of exceedance and the Design Factor are defined for each SDC in Section 2.0 of this Standard.

An SSC, when subjected to the DBE, has the greatest level of structural damage at Limit State A and the least level of structural damage at Limit State D. At Limit State A, large deformation and significant structural damage are acceptable. At Limit State D, no damage and essentially elastic behavior are the goal.

* In this Standard, the term "Seismic Design Category" has a different meaning than in the International Building Code and ASCE 7.

Limit States B and C are at intermediate levels of acceptable structural damage. The levels of acceptable structural damage defined for each Limit State are achieved in this Standard by applying the appropriate inelastic energy absorption factor, F_μ, or the deformation limits as specified in Sections 5.0 and 8.0. These design provisions include specified levels of inelastic energy absorption, structural damping, structural capacity, and material strength. These design provisions also specify that seismic analyses are to conform to ASCE 4 and provide requirements for ductile detailing.

An integral part of the seismic design criteria given in the Standard are Quality Assurance (QA) measures and independent peer review. QA measures and the involvement of peer review are expected to take place throughout the design process, beginning with establishment of the DBE and continuing through the seismic analysis and the design and detailing tasks associated with final seismic design. QA measures and peer review, as addressed in Section 9.0 of this Standard, shall follow a graded approach with increasing rigor ranging from IBC Seismic Use Group III requirements for SDC-3 to nuclear power plant requirements for SDC-5.

The overall Seismic Design Procedure for SDC 3, 4, and 5 SSCs is shown in Figure 1-1. Table 1-2 summarizes recommended earthquake design provisions for these SDCs. Specific provisions are described in detail in Sections 2.0 through 9.0 of this Standard.

Design procedures specified in this Standard conform closely to common practices. The intended users of this Standard are the civil, mechanical, and structural engineers conducting the design of nuclear facilities.

SECTION 1.2 USE OF ASCE STANDARD 43-05 WITH OTHER CODES AND STANDARDS

This Standard provides criteria for seismic design of new nuclear facilities using the concept of SDBs defined by different SDCs and Limit States associated with a graded approach.

ANSI/ANS 2.26 provides criteria for selecting SDC and Limit State that establishes the SDB for each SSC at the facility. A numerical Target Performance Goal is associated with each SDC as specified in Table 1-3. Performance goals are expressed as the mean annual probability of exceedance of the specified Limit

TABLE 1-1. SDBs for SSCs in Different SDCs and Limit States and Corresponding Seismic Design Criteria

	Limit State			
SDC	A Large Permanent Distortion (Short of Collapse)	B Moderate Permanent Distortion	C Limited Permanent Distortion	D Essentially Elastic
1	SDB-1A ASCE 7	SDB-1B ASCE 7	SDB-1C ASCE 7	SDB-1D NA
2	SDB-2A ASCE 7	SDB-2B ASCE 7	SDB-2C NA	SDB-2D NA
3	SDB-3A ASCE 43-05	SDB-3B ASCE 43-05	SDB-3C ASCE 43-05	SDB-3D ASCE 43-05
4	SDB-4A ASCE 43-05	SDB-4B ASCE 43-05	SDB-4C ASCE 43-05	SDB-4D ASCE 43-05
5	SDB-5A ASCE 43-05	SDB-5B ASCE 43-05	SDB-5C ASCE 43-05	SDB-5D ASCE 43-05

Notes:
NA = Not addressed by ASCE 7.
Shaded boxes, SDC 1 and 2, not addressed in this Standard.

TABLE 1-2. Summary of Earthquake Design Provisions

	SDC		
	3	4	5
Target Performance Goal (P_F)	1×10^{-4}	4×10^{-5}	1×10^{-5}
Probability ratio (R_P)	4	10	10
Hazard exceedence probability (H_D) ($H_D = R_P \times P_F$)	4×10^{-4}	4×10^{-4}	1×10^{-4}
DBE response spectra or time history	$DF \times UHRS$ (Section 2.0)		
Damping for structural evaluation	Section 3.4.3		
Acceptable analysis approaches for structures	ASCE 4 and Section 3.0		
Acceptable approaches for systems and components	Analysis using in-structure response spectra (ASCE 4 and Section 8.0)		
Load factors	Load factors of unity		
Inelastic energy absorption ratios	Table 5-1 and Table 8-1 used to reduce elastic response to account for permissible inelastic behavior; varies with Limit State		
Material strength	Minimum specified values		
Structural capacity	Code design strength (which includes strength reduction factors, φ, Section 4.0 and Section 8.0)		
QA program	Required within a graded approach (i.e., with increasing rigor ranging from IBC Seismic Use Group III requirements for SDC-3 to nuclear power plant requirements for SDC-5)		
Independent peer review	Required within a graded approach (i.e., with increasing rigor ranging from SDC-3 to nuclear power plant requirements for SDC-5)		

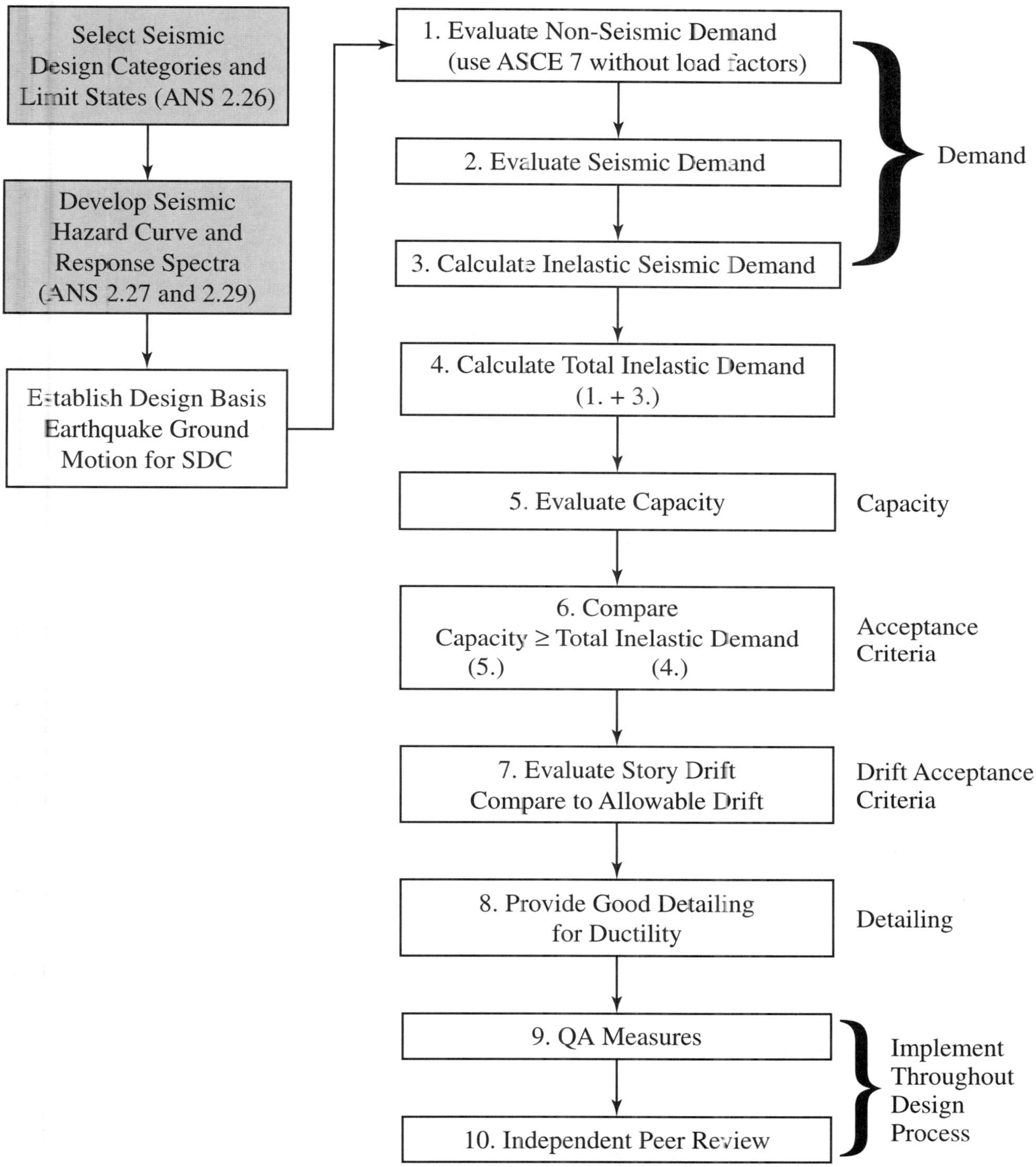

Shaded boxes represent other standards used with this Standard.

FIGURE 1-1. Seismic Design Procedure for SDC 3, 4, and 5

TABLE 1-3. Target Performance Goals for SDCs

SDC	Target Performance Goal (P_F)
1	$(< 1 \times 10^{-3})$*
2	$(< 4 \times 10^{-4})$ *
3	$\sim 1 \times 10^{-4}$
4	$\sim 4 \times 10^{-5}$
5	$\sim 1 \times 10^{-5}$

Notes:
*Working Group's assessment of performance goals approximately achieved by building codes.
Shaded boxes, SDC 1 and 2, are not addressed in this Standard.

State of structures and equipment. The deformation limits associated with each Limit State (i.e., acceptable behavior limit) are briefly described in Table 1-4.

The SDC is used to set the design earthquake levels. The Limit State is used to set the analysis methodology, design procedures, and acceptance criteria. This Standard has been developed such that the user may specify key parameters and develop a unique methodology for their application.

Five SDCs have been established in ANSI/ANS 2.26. SDCs range from those for conventional buildings to those for nuclear power plants. SSCs are assigned to SDCs relative to their importance and their inherent hazards. Conventional buildings may be assigned to SDC-1, whereas nuclear power plants may be assigned to SDC-5. ANSI/ANS 2.26 provides guidance on assigning SSCs to SDCs. In addition, four Limit States are established in ANSI/ANS 2.26. These range from near collapse to elastic behavior. The selection of SDC and Limit States for each SSC is dependent on several factors, including the overall risk of facility operation and the safety function assigned to the SSC. An SSC's SDC is based on the severity of the consequences of its failure to perform its safety function as determined by safety analysis. The seismic categorization is performed for each SSC in the facility. Hence, a facility may contain SSCs ranging from SDC-1 through SDC-5.

The following consensus codes, standards, and guidance documents shall be used, as appropriate, with this Standard:

Concrete	ACI 349, "Code requirements for nuclear safety-related concrete structures"
Steel	AISC N690, "Specification for the design, fabrication, and erection of steel safety related structures for nuclear facilities"
	AISC (LRFD), "Manual of steel construction: Load and Resistance Factor Design"
	AISC (ASD), "Manual of steel construction: Allowable stress design"
	ANSI/AISC 341-02, "Seismic provisions for structural steel buildings"
	FEMA 350, "Recommended seismic design criteria for new steel moment-frame buildings"
	Structural Engineers Association of California "The SAC steel project"
Industry Standards and Guidance	ANS 2.26, "American national standard for design categorization of nuclear facility structures, systems and components for natural phenomena hazards"
	ANS 2.27, "Site characterization requirements for natural phenomena hazards at nuclear materials facilities sites"
	ANS 2.29, "Probabilistic analysis of natural phenomena hazards for nuclear materials facilities"
	ASCE 4, "Seismic analysis of safety-related nuclear structures"
	ASCE 7, "Minimum design loads for buildings and other structures" (seismic design criteria for SDC-1 and SDC-2 SSCs)
	ASME B&PVC, "Boiler and pressure vessel code"
	ASME B31, "Piping code"
	ASME QME-1, "Qualification of active mechanical equipment used in nuclear power plants"
	IEEE 344, "Recommended practice for seismic qualification of class IE

TABLE 1-4. Structural Deformation Limits for Limit State

Limit State	Structural Deformation Limit
A	Large permanent distortion, short of collapse *Significant damage*
B	Moderate permanent distortion *Generally repairable damage*
C	Limited permanent distortion *Minimal damage*
D	Essentially elastic behavior *No damage*

	equipment for nuclear power generating stations" IEEE 628, "Standard criteria for the design, installation, and qualification of raceway systems for class 1E raceway systems for class 1E circuits for nuclear power generating stations" NFPA-13, "Installation of sprinkler systems"
Building Code Reference Documents	International Building Code (IBC) ATC 40, "Seismic evaluation and retrofit of concrete buildings" FEMA 356, "Prestandard and commentary for the seismic rehabilitation of buildings" FEMA 368, "NEHRP recommended provisions for the seismic regulations for new buildings and other structures"
Masonry	ACI 530/ASCE 5/TMS 402, "Building code requirements for masonry structures" ACI 530.1/ASCE 6/TMS-602, "Specifications for masonry structures"
Cold Formed Steel	AISI: "Specification for design of cold formed steel structural members"
Aluminum	"Aluminum association design manual"

SECTION 1.3 ALTERNATIVE METHODS TO MEET INTENT OF THIS STANDARD

The intent of this Standard is to achieve the Target Performance Goals, P_F, defined in Table 1-3, for not exceeding the Limit States defined in Table 1-4. The use of design basis ground motion as specified in Section 2.0, the criteria for evaluation of seismic demand given in Section 3.0, the criteria for structural capacity criteria given in Section 4.0, and the acceptance criteria given in Section 5.0 (as well as similar criteria for equipment and distribution systems given in Section 8.0) are all aimed at achieving the Target Performance Goals of Table 1-3.

Alternate methods are also allowed, with the provision that such methods must explicitly justify and properly incorporate into reliability calculations appropriate site-specific hazard curves, proper demands, realistic capacity estimates, and due consideration of the uncertainties in the estimation of hazard curves, demands, and capacities, and achieve the Target Performance Goals of Table 1-3.

Given the characterization of the DBE ground motion defined in Section 2.0, one such method of demonstration of compliance with this requirement (in lieu of those given in Sections 3.0, 4.0, 5.0, and 8.0, respectively) is to reasonably achieve both of the following criteria:

1. Less than about a 1% probability of unacceptable performance for the DBE ground motion defined in Section 2.0, and
2. Less than about a 10% probability of unacceptable performance for a ground motion equal to 150% of the DBE ground motion defined in Section 2.0.

SECTION 2.0 EARTHQUAKE GROUND MOTION

SECTION 2.1 SEISMIC HAZARD EVALUATION

The DBE shall be based on a Probabilistic Seismic Hazard Assessment (PSHA). Typically, the products of a PSHA are seismic hazard curves and UHRS associated with several hazard exceedance frequencies. The hazard curves and UHRS may be at the ground surface or at some specified depth.

ANSI/ANS 2.27 and 2.29 provide basic information to characterize the site and to determine the ground motion by conducting a PSHA.

Example seismic hazard curves are shown in Figure 2-1(a). Earthquake input excitation to be used for design by these provisions is defined by a mean uniform hazard response spectrum shape such as that shown in Figure 2-1(b).

SECTION 2.2 DEVELOPMENT OF DBE GROUND MOTION

2.2.1 Horizontal Ground Motion

For SDC 3, 4, and 5, a UHRS shall be specified at the mean Hazard Annual Frequency of Exceedance, H_D $(= R_P \times P_F)$, defined in Table 2-1. R_P is a design parameter needed to achieve the Target Performance Goals, P_F. The DBE ground motion shall be defined in terms of the Design Response Spectra (DRS), given by

$$DRS = DF \times UHRS \qquad \text{(Eq. 2-1)}$$

In general, the UHRS are only available at specified spectral frequencies. DF is the Design Factor, defined below, at each spectral frequency.

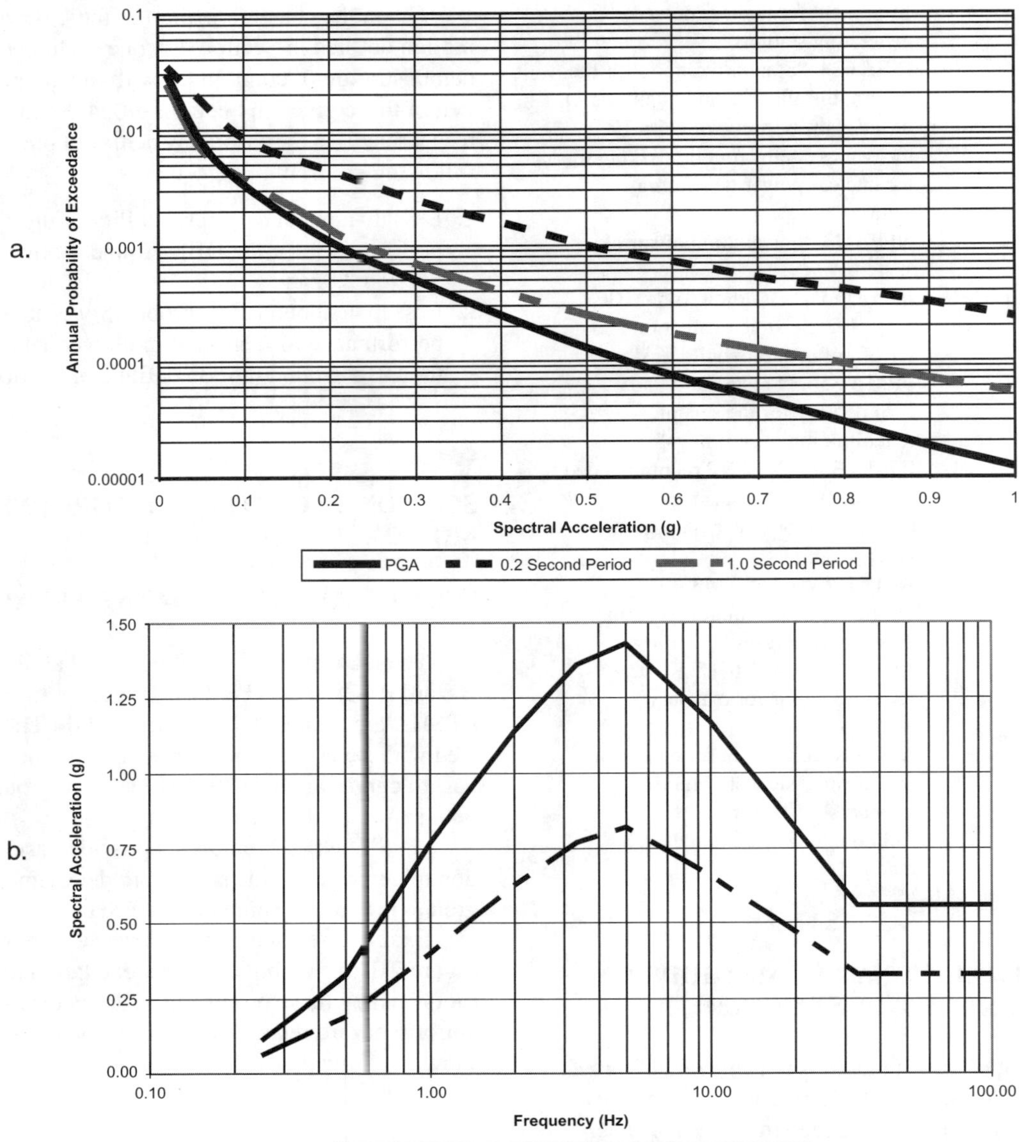

FIGURE 2-1. Earthquake Input Excitation Defined by a, Seismic Hazard Curves and b, Uniform Hazard Response Spectrum

For each spectral frequency at which the UHRS is defined, a slope factor, A_R, shall be determined from

$$A_R = \frac{SA_{0.1H_D}}{SA_{H_D}} \qquad \text{(Eq. 2-2)}$$

where SA_{H_D} is the Spectral Acceleration (SA) at the mean exceedance frequency, H_D, and $SA_{0.1H_D}$ is the SA at $0.1H_D$. Then the Design Factor, DF, at this spectral frequency is given by

$$DF = \text{Maximum}\ (DF_1, DF_2) \qquad \text{[Eq. 2-3(a)]}$$

$$DF_2 = 0.6(A_R)^{\alpha} \qquad \text{[Eq. 2-3(b)]}$$

where DF_1 and α are defined in Table 2-1.

TABLE 2-1. Design Response Spectrum Parameters

SDC	H_D	P_F	R_P	DF_1	α
3	4×10^{-4}	$\sim 1 \times 10^{-4}$	4	0.8	0.40
4	4×10^{-4}	$\sim 4 \times 10^{-5}$	10	1.0	0.80
5	1×10^{-4}	$\sim 1 \times 10^{-5}$	10	1.0	0.80

$$R_P = \frac{\text{Mean Annual Hazard Exceedance Frequency}}{} = \frac{H_D}{P_F}$$

The DRS is defined at the same control location in the free field as that at which the hazard curve and the UHRS are defined. Provisions are given in Section 2.3 for defining this DRS at other locations in the site profile.

Minimum values of DRS Peak Ground Acceleration (PGA) at the foundation level are

0.06 g for SDC-3
0.08 g for SDC-4
0.10 g for SDC-5

2.2.2 Vertical Ground Motion

Vertical ground motion shall be developed following the provisions of ASCE 4.

SECTION 2.3 METHOD TO DEFINE THE DESIGN RESPONSE SPECTRA AT VARIOUS DEPTHS IN THE SITE PROFILE

This section presents provisions for defining the DRS at other locations in the site profile besides the control location. The control location is typically defined at the bedrock outcrop. The free surface at the top of the soil profile is the most common location at which the DRS is to be determined. The DRS may also be determined at other locations in the profile following these procedures. Prior to performing the site response evaluations, the characteristic earthquakes (magnitudes and distances) at frequencies of 1 Hz and 10 Hz associated with the UHRS at the control location shall be obtained. The selection of these bounding frequencies is considered appropriate for the relatively stiff structures typical of nuclear facilities. The spectral shapes associated with these characteristic events shall then be scaled to the UHRS at 1 Hz and 10 Hz, respectively. If the envelope spectrum associated with these two scaled spectra does not fall more than 10% below the UHRS at any frequency in the frequency range of interest, site response evaluations can be performed for these two bedrock spectra. If the envelope spectrum does fall below the UHRS at some intermediate frequency, a third intermediate spectral shape shall be determined using the characteristic event appropriate for that intermediate frequency. This spectral shape shall then be scaled to the UHRS at the intermediate frequency.

The approach for obtaining the DRS at the ground surface (or at some other intermediate depth) is summarized as follows:

(a) Convolve the UHRS at the hazard mean annual exceedance probability, H_D, at depth to obtain the corresponding UHRS at H_D at the ground surface (or other location in the soil column) using site-specific soil properties.
(b) Convolve the UHRS at $0.1H_D$ at depth to obtain the corresponding UHRS at $0.1H_D$ at the surface (or other location).
(c) Determine the slope factor, A_R, from the ratio of $UHRS_{0.1H_D}/UHRS_{H_D}$ at the ground surface, computed over the spectral frequency range, frequency by frequency, using Eq. (2-2).
(d) Use Eq. (2-3) to develop the Design Factor, DF, at each spectral frequency, at the ground surface.
(e) Modify the $UHRS_{H_D}$ at the surface with DF to obtain the DRS at the ground surface.

The number of convolution calculations performed must be sufficient to capture effects of the variability and uncertainty in soil properties on site response.

SECTION 2.4 CRITERIA FOR DEVELOPING SYNTHETIC OR MODIFIED RECORDED TIME HISTORIES

Ground motions that are generated to "match" or "envelop" given design response spectral shapes defined in Section 2.2 shall comply with steps (a) through (f) below. The general objective is to generate a modified recorded or synthetic accelerogram that achieves approximately a mean-based fit to the target spectrum; that is, the average ratio of the spectral acceleration calculated from the accelerogram to the target, where the ratio is calculated frequency by frequency, is only slightly greater than one. The aim is to achieve an accelerogram that does not have significant gaps in the Fourier amplitude spectrum, but which is not biased high with respect to the target. Records biased high with respect to a spectral target may overdrive (overestimate damping and stiffness reduction) a site soil column or structure when nonlinear effects are important.

(a) The time history shall have a sufficiently small time increment and sufficiently long duration. Records shall have a Nyquist frequency of at least 50 Hz (e.g., a time increment of at most 0.010 s) and a total duration of at least 20 s. If frequencies higher than 50 Hz are of interest, the time increment of the record must be suitably reduced to provide a Nyquist frequency ($N_y = 1/(2\ \Delta t)$, where Δt = time increment) above the maximum frequency of interest. The total duration of the record can be increased by zero packing to satisfy these frequency criteria.

(b) Spectral accelerations at 5% damping shall be computed at a minimum of 100 points per frequency decade, uniformly spaced over the log frequency scale from 0.1 Hz to 50 Hz or the Nyquist frequency. If the target response spectrum is defined in the frequency range from 0.2 Hz to 25 Hz, the comparison of the synthetic motion response spectrum with the target spectrum shall be made at each frequency computed in this frequency range.

(c) The computed 5% damped response spectrum of the accelerogram (if one synthetic motion is used for analysis) or of the average of all accelerograms (if a suite of motions is used for analysis) shall not fall more than 10% below the target spectrum at any one frequency. To prevent spectra in large frequency windows from falling below the target spectrum, the spectra within a frequency window of no larger than ±10% centered on the frequency shall be allowed to fall below the target spectrum. This corresponds to spectra at no more than nine adjacent frequency points defined in (b) above from falling below the target spectrum.

(d) In lieu of the power spectral density requirement of ASCE 4, the computed 5% damped response spectrum of the synthetic ground motion (if one synthetic motion is used for analysis) or the mean of the 5% damped response spectra (if a suite of motions is used for analysis) shall not exceed the target spectrum at any frequency by more than 30% (a factor of 1.3) in the frequency range between 0.2 Hz and 25 Hz. If the spectrum for the accelerogram exceeds the target spectrum by more than 30% at any frequency in this frequency range, the power spectral density of the accelerogram needs to be computed and shown to not have significant gaps in energy at any frequency over this frequency range.

(e) Because of the high variability in time domain characteristics of recorded earthquakes of similar magnitudes and at similar distances, strict time domain criteria are not recommended. However, synthetic motions defined as described above shall have strong motion durations (defined by the 5% to 75% Arias intensity), and ratios V/A and AD/V^2 (A, V, and D are the peak ground acceleration, ground velocity, and ground displacement, respectively), which are generally consistent with characteristic values for the magnitude and distance of the appropriate controlling events defined for the UHRS.

(f) To be considered statistically independent, the directional correlation coefficients between pairs of records shall not exceed a value of 0.30 (see Definitions in this Standard). Simply shifting the starting time of a given accelerogram does not constitute the establishment of a different accelerogram. If uncoupled response of the structure is expected, then only one time history is required. Then, the seismic analysis for each direction can be performed separately and then combined by the square root of the sum of the squares (SRSS).

Synthetic, recorded, or modified recorded earthquake ground motion time histories may be used for linear seismic analyses. Actual recorded earthquake ground motion or modified recorded ground motion shall be used for nonlinear seismic analyses. For nonlinear analyses, it is desirable to utilize actual recorded earthquake ground motion. However, to meet the requirements of steps (a) through (f) above, as many as 30 recorded earthquake motions would be required. As a result, it is acceptable to use modified recorded earthquake accelerograms that shall meet steps (a) through (f). A modified recorded accelerogram is a time-history record of acceleration versus time that has been produced from an actual recorded earthquake time history. However, the Fourier amplitudes are scaled such that the resulting response spectrum envelops the target response spectrum in the manner described above. The Fourier phasing from the recorded earthquake time history is preserved in a modified recorded earthquake accelerogram.

The selection of recorded or modified recorded accelerograms is based on the identification of dominant magnitude/distance pairs that impact the site DRS. The accelerograms used for nonlinear seismic calculations shall be selected from the appropriate magnitude/distance (M/D) bins. It may be necessary to produce different accelerograms that characterize the seismic hazard at appropriate low (about 1 Hz) and high (10 Hz) frequency. Alternatively, the accelerograms may be selected to match the dominant M/D pairs at the peak velocity and acceleration segments of the design spectrum. If behavior at the peak displacement frequency range is of interest, additional accelerograms may be selected whose controlling event is appropriate at these frequencies.

If recorded accelerograms are used directly as input to the nonlinear analyses, the suite of time histories shall meet the requirements of steps (a) through (f) above. If modified recorded time histories are generated to match the target spectrum, it is important to ensure that the phase spectra of the motions are generated from recorded motions in the appropriate *M/D* bins. In addition, the strong motion duration (as defined as the duration from the 5% to 75% Arias intensity) shall fall within the range appropriate for the *M/D* bin. In accepting the suite of motions, the range in variation in rise time of the Arias intensity shall be considered, such that all do not have the same rise time characteristics.

SECTION 3.0 EVALUATION OF SEISMIC DEMAND

SECTION 3.1 INTRODUCTION

Seismic demand shall be computed in accordance with the requirements of ASCE 4. Seismic demand shall be computed using linear equivalent static analysis, linear dynamic analysis, complex frequency response methods, or nonlinear analysis in accordance with the following sections and ASCE 4. Regardless of the procedure followed, it is important that

1. The input to the SSC be defined by either a DRS (Section 2.2) or a response spectrum compatible acceleration time history (Section 2.4).
2. The important natural frequencies of the SSC be estimated, or that the peak of the design spectrum, multiplied by an appropriate factor (Section 3.2.1),* be used as input. Soil–structure interaction and multimode effects shall be considered.
3. A load path evaluation for seismic induced inertial forces be performed. A continuous load path, or paths, with adequate strength and stiffness shall be provided to transfer all forces from the point of application to the foundation.
4. Seismic demand shall be obtained for the three orthogonal (two horizontal and one vertical) components of earthquake motion in accordance with ASCE 4. In general, the orthogonal axes shall be aligned with the principal axes of the structure.
5. All vertical load path elements shall be designed for the lateral displacements induced by seismic loads on the structure.

* From ASCE 4-98, Section 3.2.5.

SECTION 3.2 LINEAR ANALYSIS

3.2.1 Linear Equivalent-Static Analysis

An equivalent-static analysis may be used to evaluate single-point-of-attachment cantilever models with essentially uniform mass distribution, or other simple structures that can be idealized as a single-degree-of-freedom system. For cantilever models with essentially uniform mass distribution, the equivalent-static load base shear shall be determined by multiplying the cantilevered structure, equipment, or component masses by an acceleration equal to the peak of the input response spectrum. For these structures, the base moment shall be determined by using an acceleration equal to 1.1 times the peak of the applicable response spectrum. The resulting load shall be applied at the center of gravity of the structure.

For cantilevers with nonuniform mass distribution and other simple multiple–degree-of-freedom structures in which the predominant or fundamental mode shape of the structure has a curvature in one direction only (similar to a cantilever mode), the equivalent-static load shall be determined by multiplying the structure, equipment, or component masses by an acceleration equal to 1.5 times the peak acceleration of the applicable response spectrum. A smaller factor may be used, if justified.

Alternately, the spectral acceleration value at the fundamental frequency of the structure may be used if a modal solution has been obtained in accordance with ASCE 4. The use of the 1.1 or 1.5 factors defined above shall be applied to the spectral acceleration value determined at the fundamental frequency.

3.2.2 Linear Dynamic Analysis

Linear dynamic analysis may be used for any structure and may be performed using either response-spectrum or time-history approaches. Time-history approaches may use either direct integration or modal superposition methods in accordance with Section 3.2.2 of ASCE 4. P-Δ effects shall be included, if significant. If inclusion of P-Δ effects results in greater than a 10% increase in the imposed moment demand on a structural member, the effects shall be included; otherwise, they may be omitted.

SECTION 3.3 NONLINEAR ANALYSIS

Nonlinear seismic response analysis may need to be performed when significant nonlinear behavior is expected in some elements or when significant irregularities exist. This method requires definition of the load-deformation behavior of individual elements or the overall structural system. The nonlinear load-deformation curves used in analysis shall reflect behavior based on experimental data, which may be approximated by linear or curved segments Nonlinear behavior shall be determined under monotonically increasing lateral deformation when nonlinear static analysis (pushover analysis) is performed. In the case of nonlinear dynamic analysis, appropriate load-deformation curves under multiple reversed deformation cycles shall be used.

3.3.1 Nonlinear Static Analysis

Structures whose response is dominated by a single mode may be evaluated using a nonlinear equivalent-static (pushover) analysis, provided that an effective frequency and damping are used to quantify the nonlinear response. Nonlinear equivalent-static methods of analysis shall follow the guidance provided in FEMA-356 for the target displacement method or in ATC-40 for the capacity spectrum method.

3.3.2 Nonlinear Dynamic Analysis

Nonlinear dynamic procedures shall follow the guidance provided in Section 3.2 of ASCE 4. Nonlinear dynamic analysis shall

- Have sufficient degrees of freedom to represent important responses of the structure. Single-degree-of-freedom models may be used for structures whose response is dominated by a single mode.
- Include P-Δ forces, if significant.
- Appropriately represent both the monotonic (backbone) and cyclic behavior of nonlinear elements. Members that exhibit pinched hysteretic behavior in laboratory tests shall be represented in the analysis with elements that represent similar pinching characteristics. Mean force-deflection properties shall be used.
- Approximate plastic hinge lengths for frame members by one beam depth, developed by rational analysis, or justified by comparison to test data.

When performing such nonlinear calculations, at least three different modified recorded accelerograms shall be used to determine potential nonlinear response. If less than five accelerograms are used, the largest response shall be used in making demand-to-capacity checks. If five or more accelerograms are used, the mean of the calculated responses may be used in making demand-to-capacity checks. If design spectrum matching is done separately for the low-frequency (about 1 Hz) and high-frequency (about 10 Hz) ranges, then at least three time histories are required for each frequency range.

SECTION 3.4 MODELING AND INPUT PARAMETERS

Modeling of SSCs for seismic analysis shall follow Section 3.1 of ASCE 4.

3.4.1 Effective Stiffness of Reinforced Concrete Members

In lieu of a detailed stiffness calculation, the effective stiffness of reinforced concrete members provided in Table 3-1 shall be used in linear elastic static or dynamic analysis. When finite element methods are used, the element stiffness shall be modified using the effective stiffness factor for the dominant response parameter.

3.4.2 Mass

The mathematical model used for determining seismic response shall include mass due to the following:

- Weight of the structure
- Weight of permanent equipment
- Expected live load, not less than 25% of the specified design live loads

Design snow loads of 30 psf or less need not be included. Where snow loads exceed 30 psf, the design snow load shall be included, but it may be reduced up to 75% where consideration of siting, configuration, and load duration warrant.

3.4.3 Damping Values for SSCs

Damping values to be used in linear elastic analyses for determining seismic design loads for SSCs are presented in Table 3-2 as a function of the average Response Level in the seismic load-resisting elements represented by the demand-to-capacity ratio (D_e/C). The D_e/C ratios are calculated on an element basis (C = code capacity, D_e = total elastic demand, including non-seismic loads). The appropriate Response Level can be estimated from Table 3-3.

Response Level 3 damping may be used for evaluating seismic-induced forces and moments in structural members by elastic analysis without consideration of the actual Response Level for Limit States A, B, or C. Response Level 2 damping may be used for Limit State D.

Consideration of the actual Response Level is required for generation of in-structure response spectra. In lieu of iterative analyses to determine the actual Response Level and associated damping value, Response Level 1 damping values may be used for generation of in-structure spectra. Response Level 1 damping values must be used if elastic buckling considerations control the design.

TABLE 3-1. Effective Stiffness of Reinforced Concrete Members

Member	Flexural Rigidity	Shear Rigidity	Axial Rigidity
Beams—Nonprestressed	$0.5\ E_cI_g$	G_cA_W	
Beams—Prestressed	E_cI_g	G_cA_W	
Columns in compression	$0.7\ E_cI_g$	G_cA_W	E_cA_g
Columns in tension	$0.5\ E_cI_g$	G_cA_W	E_sA_s
Walls and diaphragms—Uncracked	E_cI_g $(f_b < f_{cr})$	G_cA_W $(V < V_c)$	E_cA_g
Walls and diaphragms—Cracked	$0.5\ E_cI_g$ $(f_b > f_{cr})$	$0.5\ G_cA_W$ $(V > V_c)$	E_cA_g

Notes:
A_g = Gross area of the concrete section
A_s = Gross area of the reinforcing steel
A_w = Web area
E_c = Concrete compressive modulus, from ACI-349 $57{,}000(f'_c)^{1/2}$
E_s = Steel modulus
f_b = Bending stress
f_{cr} = Cracking stress
G_c = Concrete shear modulus = $0.4\ E_c$
I_g = Gross moment of inertia
V = Wall shear
V_c = Nominal concrete shear capacity

TABLE 3-2. Specified Damping Values for Dynamic Analysis

	Damping (% of Critical)		
Type of Component	Response Level 1	Response Level 2	Response Level 3
Welded and friction-bolted metal structures	2	4	7
Bearing-bolted metal structures	4	7	10
Prestressed concrete structures (without complete loss of prestress)	2	5	7
Reinforced concrete structures	4	7	10
Reinforced masonry shear walls	4	7	10
Piping	5	5	5
Distribution systems:			
• Cable trays 50% or more full and in-structure response spectrum Zero Period Acceleration of 0.25 g or greater	5	10	15
• For other cable trays, cable trays with rigid fireproofing and conduits	5	7	7
Massive, low-stressed mechanical components (pumps, compressors, fans, motors, etc.)	2	3	—*
Light welded instrument racks	2	3	—*
Electrical cabinets and other equipment	3	4	5**
Liquid containing metal tanks:			
• Impulsive mode	2	3	4
• Sloshing mode	0.5	0.5	0.5

Notes:
* Should not be stressed to Response Level 3. Use damping for Response Level 2.
** May be used for anchorage and structural failure modes that are accompanied by at least some inelastic response. Response Level 1 damping values shall be used for functional failure modes such as relay chatter or relative displacement issues that may occur at a low cabinet stress level.

TABLE 3-3. Estimating Damping Response Level

Response Level*	D_e/C
1	≤ 0.5
2	≈ 0.5 to 1.0
3	≥ 1.0

Note:
* Consideration of these damping levels is required only in the generation of floor or amplified response spectra to be used as input to subcomponents mounted on the supporting structure. For analysis of structures including soil–structure interaction effects, D_e/C ratios for the best estimate case of soil properties shall be used to determine Response Level.

If a nonlinear inelastic response analysis is performed that explicitly incorporates the hysteretic energy dissipation, Response Level 1 damping values shall be used to avoid the double counting of the hysteretic energy dissipation, which would result from the use of higher damping values. Response Level 2 values may be used if they can be justified. A summary of the maximum Response Level that shall be used for selecting the damping value is shown in Table 3-4.

SECTION 4.0 EVALUATION OF STRUCTURAL CAPACITY

SECTION 4.1 STRUCTURAL SYSTEMS

4.1.1 Definitions

Structural systems shall be classified as one of the types listed in Table 4-1 and defined in this section. Many buildings have a single structural system; for example, storage vaults may be composed of a building frame consisting of reinforced concrete shear walls. Also, many structures may be composed of a number of different structural systems. A structure may also have different structural systems (e.g., braced frame and shear wall) in each of its major horizontal dimensions.

BEARING WALL SYSTEM. A structural system without a complete vertical load-carrying space frame. Bearing walls or bracing systems provide support for some of the gravity loads. Resistance to lateral load is provided by shear walls or braced frames, which also provide resistance to gravity loads

BUILDING FRAME SYSTEM. A structural system with an essentially complete space frame providing support for gravity loads. Resistance to lateral load is provided by shear walls or braced frames.

MOMENT-RESISTING FRAME SYSTEM. A structural system with an essentially complete space frame providing support for gravity loads. Moment-resisting frames provide resistance to lateral load primarily by flexural action of members.

DUAL SYSTEM. A structural system with the following features:

- An essentially complete space frame that provides support for gravity loads.
- Lateral load resistance that is provided by shear walls or braced frames and special moment-resisting frames. The moment-resisting frames shall be designed to independently resist at least 25% of the design base shear.
- Total design base shear resistance. The two systems shall be designed to resist the total design base shear in proportion to their relative rigidities considering the interaction of the dual system at all levels.

NONBUILDING STRUCTURAL SYSTEM. Nonbuilding structures include all self-supporting

TABLE 3-4. Summary of Maximum Response Level for Damping

Elastic buckling conditions control design	Response Level 1
Generation of in-structure spectra	Response Level 1 (Response Level 2, if justified)
Limit State D	Response Level 2
Limit States A, B, or C:	
Elastic analysis	Response Level 3*
Inelastic time-history response analysis	Response Level 1 (Response Level 2, if justified)

Note:
* Only to be used with adequate ductile detailing. However, functionality of SSCs must be given due consideration. Response Level 3 corresponds to Limit State C; Response Level 3 may also be used for Limit States A and B.

TABLE 4-1. Acceptable Structural Systems for use in Nuclear Facilities

Basic Structural System	Lateral Force-Resisting System Description
Bearing wall system	1. Shear walls:
	a. Concrete with special detailing
	b. Masonry with special (Category D) detailing
	2. Light steel-framed bearing walls with tension-only bracing
	3. Braced frames where bracing carries gravity load:
	a. Steel eccentrically braced frame (EBF)
	b. Special and ordinary steel concentrically braced frame
Building frame system	Steel EBF
	Shear walls:
	a. Concrete with special detailing
	b. Masonry with special (Category D) detailing
	Special and ordinary steel concentrically braced frames
Moment-resisting frame system	Special moment-resisting frame (SMRF):
	a. Steel
	b. Concrete
Dual system*	Shear walls:
	a. Concrete with SMRF
	b. Masonry with SMRF
	Steel EBF:
	a. With steel SMRF
	Special concentrically braced frames
	a. Steel with steel SMRF

Note:
*Dual systems are permitted provided the primary lateral load-resisting system is designed to take 100% of the lateral design loads in each direction. The secondary system should be designed to take 25% of the lateral design loads. No additional credit is given for the use of dual systems in the manner of increased inelastic energy absorption factors.

structures other than buildings that carry gravity loads and resist the effects of earthquakes. Nonbuilding structures shall be designed to provide the strength and stiffness required to resist the minimum lateral forces specified in this Standard. Design shall conform to the applicable provisions of other sections of this Standard and by site-specific requirements.

USE LIMITS. Certain structural systems shall not be used for SDC 3, 4, and 5, as discussed in Section 4.1.3.

4.1.2 Acceptable Structural Systems for Nuclear Facilities

These criteria have adapted the guidance of nationally accepted consensus codes (e.g., the IBC) where possible. The national consensus codes, however, are intended for application to a larger population of structure types than those that are typically used for nuclear facilities. Nuclear structures generally perform a confinement function such that the integrity of the exterior walls, which act as particulate filtration boundaries for accidents beyond a design basis, is desired and frequently required. This does not imply that the exterior walls of a nuclear facility will remain leak-tight following a DBE. In fact, undamaged concrete has intrinsic air permeability from cracking due to normal temperature and shrinkage stresses. This does imply, however, that the damaged state of an exterior wall designed as a confinement element in a nuclear facility will be limited to tight cracks that serve as particulate filters.

Preferred structural systems for nuclear facilities possess adequate strength and ductility to survive the DBE. Preferred structural systems that provide confinement also possess adequate stiffness to limit interstory deformations such that cracking in exterior walls is minimized. Structural systems that possess these qualities include the following:

1. Structures that rely on specially reinforced concrete or masonry shear walls providing the lateral-force–resisting system
2. Slab/wall frames

Specific criteria for steel braced frames, steel moment frames, concrete moment frames, and concrete shear walls are provided in this Standard. Other systems, listed in Table 4-1, may also be used for nuclear facilities; however, no additional guidance is provided for these systems. For these other systems or materials, national consensus standards shall be used for design, supplemented by the additional requirements for seismic demand, seismic capacity, ductile detailing, acceptance criteria, and QA/independent peer review specified in this Standard.

4.1.3 Prohibited Structural Systems

The national consensus codes (e.g., IBC) permit the use of some systems that are not appropriate for use in nuclear facilities. These systems are unacceptable due to either large interstory drifts at high seismic demands or brittle failure mechanisms. Structural systems specifically prohibited for use in the design of the lateral-force–resisting system of nuclear facilities include the following:

- Ordinary moment-resisting frame systems
- Intermediate moment-resisting frame systems
- Plain concrete systems
- Precast concrete systems, which use gravity-only bearing connections
- Unreinforced masonry systems
- Wood structures

The use of other systems listed in NEHRP design provisions or in the IBC must be justified. The justification shall include, but is not limited to, a discussion of the performance of systems in earthquakes, a description of the expected performance of the system at the design basis ground motion, and a discussion of the building's ability to provide adequate confinement following the DBE.

SECTION 4.2 STRUCTURAL CAPACITIES

4.2.1 General

Structural capacities shall be based on the strength design approach or allowable stress design levels amplified to strength design amplitudes. Either approach may be used in the design process as described in the following paragraphs.

In the strength design approach, the nominal capacities of structural members shall be based on the applicable code provisions. The code equations are generally based on the specified minimum compressive strength or yield strength of the material. The specified strengths shall be verified by testing (i.e., concrete cylinder tests or steel coupon tests) during production. An adequate number of samples in accordance with material specifications shall be used.

The code strength is then determined by applying a capacity reduction factor from the appropriate design code to the nominal capacity. Thus, the code capacity is defined as

$$C = \phi C_N \qquad \text{(Eq. 4-1)}$$

where

C = Code design capacity
ϕ = Capacity reduction factor (strength reduction factor)
C_N = Nominal code capacity

Alternatively, the equivalent code capacity may be defined using the conventional allowable stresses amplified by a capacity increase factor:

$$C = KC_w \qquad \text{(Eq. 4-2)}$$

where

K = Capacity increase factor as given in Table 4-2
C_w = Capacity based on allowable stresses without the one-third increase for seismic stresses

4.2.2 Reinforced Concrete

Design of concrete structures shall be based on the strength design method. The nominal capacity of reinforced concrete members shall be calculated in accordance with ACI 349. The code capacity of the reinforced concrete members shall be determined by multiplying the nominal capacity by the capacity reduction factors given in ACI 349.

The capacity of low-rise shear walls as prescribed in ACI 349 may be too conservative. In lieu of the

TABLE 4-2. Conversion Between ASD and Strength-Based (LRFD) Capacities

Component	ASD *FS*	ϕ	$K = C_{LRFD}/C_{ASD}$
Bending and shear	1.5 to 1.67	0.9	1.5 to 1.34; use 1.4
Axial compression	1.67 to 1.92	0.85	1.62 to 1.41, use 1.5
Axial tension:			
Yield stress	1.67	0.9	1.5
Ultimate stress	2	0.75	1.5

code provisions, the low-rise shear wall capacity may be established using the procedure given in Section 4.2.3.

4.2.3 Capacity of Low-Rise Concrete Shear Walls

In lieu of the ACI 349 code provisions, the shear strength of low-rise ($h_w/l_w \leq 2.0$) concrete walls with boundary elements or end walls can be calculated using the following equation:

$$v_u = \phi\left[8.3\sqrt{f'_c} - 3.4\sqrt{f'_c}\left(\frac{h_w}{l_w} - 0.5\right) + \frac{N_A}{4l_w t_n} + \rho_{se} f_y\right] \quad \text{(Eq. 4-3)}$$

where

ϕ = Capacity reduction factor (= 0.8)
v_u = Ultimate shear strength (psi)
f'_c = Concrete compressive strength (psi)
h_w = Wall height (in.)
l_w = Wall length (in.)
N_A = Axial load (lb)
t_n = Wall thickness (in.)
$\rho_{se} = A\rho_v + B\rho_u$
f_y = Steel yield strength (psi)
ρ_v = Vertical steel reinforcement ratio
ρ_u = Horizontal steel reinforcement ratio
A, B = Constants given as follows:

$h_w/l_w \leq 0.5$	$A = 1$	$B = 0$
$0.5 \leq h_w/l_w \leq 1.5$	$A = -h_w/l_w + 1.5$	$B = h_w/l_w - 0.5$
$h_w/l_w \geq 1.5$	$A = 0$	$B = 1$

This equation is applicable for shear walls with h_w/l_w less than or equal to 2.0 and for horizontal and vertical steel reinforcement ratios less than or equal to 0.01. For walls with h_w/l_w greater than 2.0, the shear wall provisions of the ACI 349 code shall be used. If reinforcement ratios exceed 0.01, ρ_{se} in Eq. (4-3) shall be limited to 0.01. In no case shall v_u exceed $20\ \phi (f'_c)^{1/2}$.

The total shear capacity is

$$V_u = v_u\, d\, t_n \quad \text{(Eq. 4-4)}$$

where d is the distance from the extreme compression fiber to the center of force of all reinforcement in tension, which may be determined from a strain compatibility analysis. If such an analysis is not performed, then

$$d = 0.6\, l_w \quad \text{(Eq. 4-5)}$$

The analyst shall also check other failure modes, which include concrete web compression capacity and flexural capacity, following the procedures given in the ACI 349 code. For lightly reinforced walls (i.e., steel reinforcement ratios less than 0.01), such as generally occur in nuclear power plants and other heavy industrial facilities, compression failure is not likely to control. For a low-rise shear wall, flexural capacity seldom governs. Requirements for boundary elements given in the ACI 349 code shall be checked.

4.2.4 Structural Steel

Design of steel structures shall be performed using any of the following three methods. The selected method must be used in its entirety.

- **AISC/LRFD.** The nominal strength of structural steel members shall be calculated in accordance with the American Institute of Steel Construction (AISC) Load and Resistance Factor Design (LRFD) Manual of Steel Construction, as modified by the AISC Seismic Provisions. The code strength of steel members shall be determined by multiplying the nominal capacity by the capacity reduction factors provided by the AISC LRFD manual.
- **AISC/ASD.** The allowable stresses of structural steel members in accordance with the AISC Allowable Stress Design (ASD) Manual of Steel Construction, as modified by the AISC Seismic Provisions. The allowable stress capacities are scaled to the LRFD strength-based nominal code capacities by

$$C_{LRFD} = \phi\, FS\, C_{ASD}$$

where ϕ is the LRFD strength reduction factor and FS is the ASD factor of safety for a particular component. Note that the allowable stress capacities, C_{ASD}, given by AISC are F_y/FS or F_u/FS and that the one-third stress increase has been omitted. Thus, the LRFD-to-allowable-stress conversions are summarized in Table 4-2.

- **AISC/N690.** Alternately, stress limit coefficients in AISC N690 and its Supplement No. 1, appropriate for load combinations that include the Safe Shutdown Earthquake (SSE), can be used to scale the AISC/ASD-allowable stresses to LRFD strength-based nominal code capacities. AISC N690 shall be modified by the ANSI/AISC 341-02 Provisions, where appropriate.

4.2.5 Reinforced Masonry

Design of reinforced masonry structures shall be based on strength design methods. In the strength method, the nominal capacity of reinforced masonry elements shall be calculated in accordance with the IBC, Section 2108. The strength reduction factors for determining the code capacity are provided in the same section. Alternatively, strength design of masonry structures may be based on the "Limit States design" approach of ACI 530, Building Code Requirements for Masonry Structures.

SECTION 4.3 DEFORMATION AND ROTATION CAPACITIES

Deformation and rotation capacities of structural elements shall be based on test data. The limits given in Section 5.0 are based on previous test data on structural elements encountered in typical building design. These limits are applicable to the design of structural elements that are within the scope of test parameters.

For designs that are outside the test parameters, a suitable test program may be necessary to establish the specific deformation and rotation capacities. Alternatively, the design may be revised to bring all elements within the boundaries established by previous test data.

SECTION 5.0 LOAD COMBINATIONS AND ACCEPTANCE CRITERIA FOR STRUCTURES

SECTION 5.1 LOAD COMBINATIONS

5.1.1 General

The structure shall be proportioned and detailed to resist the effects of gravity loads, operating loads, earthquake loads, other natural phenomena loads, and applicable accident loads. The loading combination for seismic loads is defined below, while other load combinations are to be defined in the project design criteria.

5.1.2 Seismic Loading Combinations

Seismic load combinations for strength-based acceptance criteria shall consider the energy dissipation factor, F_{μ}. Seismic loading combinations for deformation-based acceptance criteria shall combine the unfactored effects of both seismic and non-seismic loads. Details of both approaches are given below.

5.1.2.1 Seismic Loading Combinations for Strength-Based Acceptance Criteria

For elastic analyses, the total demand acting on an element shall be the sum of non-seismic demand, D_{NS}, and seismic demand, D_S, per the following load combination, as appropriate:

For bending moment, in-plane shear, and axial load in diagonal bracing, use

$$D = D_{NS} + \frac{D_S}{F_{\mu S}} \qquad \text{[Eq. 5-1(a)]}$$

For other axial loads, other shear loads, and torsion, use

$$D = D_{NS} + \frac{D_S}{1.0} \qquad \text{[Eq. 5-1(b)]}$$

where

D = Total demand acting on an element.

D_{NS} = Non-seismic demand acting on an element. Non-seismic demand shall include the mean effects of dead, live, equipment, fluid, snow, and at-rest lateral soil loads.

D_S = Calculated seismic response to the DBE using an elastic analysis approach (by either response spectrum or time-history analysis) discussed in Section 3.0 and the appropriate damping values from Table 3-2.

$F_{\mu S}$ = System inelastic energy absorption factor for structural elements.

$F_{\mu S}$ is determined from the inelastic energy absorption factor, F_{μ}, provided in Table 5-1, but it may be reduced from the tabular values because the structure has a weak or a soft story or because the predominate structural response occurs at a frequency greater than the amplified acceleration region of the DRS. A weak story is one in which the story lateral strength is less than 80% of the story above. The story strength is the total strength of seismic-resisting elements sharing the story shear for the direction under consideration. A soft story is one in which the lateral stiffness is less than 70% of that in the story above or less than 80% of the average stiffness of the three stories above.

To account for reduced inelastic energy absorption capacity of a structure with a weak or soft story, $F_{\mu S1}$ is determined from the following:

When a structure has either a weak story or soft story,

$$F_{\mu S1} = 1 + 2(F_{\mu} - 1)\frac{n - k + 1}{n(n + 1)} \qquad \text{[Eq. 5-2(a)]}$$

If the structure does not have either a weak or soft story,

$$F_{\mu S1} = F_{\mu} \qquad \text{[Eq. 5-2(b)]}$$

TABLE 5-1. Inelastic Energy Absorption Factor, F_μ

Limit State	Reduction Factor, F_μ*		
	LS-A	LS-B	LS-C
SMRF reinforced concrete moment frames			
Beams $(15 \leq \ell/h)$	5.25	4.0	2.5
Beams $(\ell/h \leq 10)$	3.25	3.0	2.5
Columns**	2.0	1.75	1.5
Reinforced concrete shear wall, in plane:			
Bending controlled walls, $\frac{h_w}{\ell_w} \geq 2.0$			
$6\sqrt{f'_c} < f_v$	2.25	2.0	1.75
$f_v < 3\sqrt{f'_c}$	2.5	2.25	1.75
Shear controlled walls, $\frac{h_w}{\ell_w} < 2.0$	2.0	1.75	1.5
SMRF steel moment frames (bending only):			
Beams and columns** $P < 0.2\,P_y$	5.25	3.5	2.5
Columns** $P = 0.3\,P_y$	4.25	3.0	1.25
Columns** $P = 0.4\,P_y$	3.25	2.25	1.25
Columns** $P = 0.5\,P_y$	1.75	1.5	1.25
Columns** $P > 0.5\,P_y$	1.0	1.0	1.0
Steel braced frames:			
Special concentric (bracing members)	4	3	2
Ordinary concentric (bracing members)	2.5	2.0	1.5
Eccentric			
Short link beam $1.6\frac{Z}{0.6A_w} < e$	6	4	2
Long link beam $e < 2.6\frac{Z}{0.6A_w}$	5	3.5	2.5
Chevron bracing	2.5	2.0	1.5
Slab/wall moment frames:			
Roof slabs, floor slabs, beams and walls of reinforced concrete			
$(15 \leq \ell/h)$	2.5	2.25	2.0
$(\ell/h \leq 10)$	2.25	2.0	2.0

Notes:
* $F_\mu = 1.0$ for Limit State D.
** F_μ for compression and shear in columns shall be unity (Eqs. [5-1(a)] and [5-1(b)]
A_w = Web area of a link beam
e = Eccentricity measured along the link beam
f'_c = Concrete compressive strength
f_v = Average shear stress, which is equal to shear divided by the shear area
h_w = Height of the entire shear wall or segment of the shear wall considered
P = Nominal axial load
P_y = Axial yield strength
Z = Link beam plastic section modulus
ℓ = Span length
h = Beam depth
ℓ_w = Length of shear wall or segment of shear wall considered in direction of shear force

where

F_μ = Inelastic energy absorption factor discussed in Section 5.1.2.3
n = Number of stories in the structure
k = Story level of the highest weak/soft story in the structure

Judgment should be used in determining the number of stories in facilities that have irregular geometry, discontinuous floor elevations, and partial height walls.

The adjustment for the effects of a weak/soft story is applied only to element seismic demands in the weak/soft story and all stories below. For stories above the weak/soft story, $F_{\mu S1}$ equal to F_μ shall be used.

To account for reduced inelastic energy absorption capacity of a structure with predominant structural response at a frequency greater than the amplified acceleration region of the DRS, $F_{\mu S}$ is determined from

$$F_{\mu S} = F_{\mu S1}\left(\frac{SA(f_n)}{SA(f_e)}\right) \geq 1.0 \qquad \text{(Eq. 5-3)}$$

where

$SA(f_n)$ = Spectral acceleration at the frequency of the predominant structural mode, f_n
$SA(f_e)$ = Spectral acceleration at an effective natural frequency for nonlinear seismic response, f_e, as defined by Eq. (5-4)

When $f_n \leq f_{PEAK\ SA}$ then $f_e = f_n$
When $f_n > f_{PEAK\ SA}$ then

$$f_e = f_n\sqrt{\frac{2}{F_{\mu S1}^2 + 1}} \geq f_{PEAK\ SA} \qquad \text{(Eq. 5-4)}$$

where $f_{PEAK\ SA}$ is the upper frequency of the amplified acceleration region of the DRS.

5.1.2.2 *Seismic Loading Combinations for Displacement-Based Acceptance Criteria*

The total demand acting on an element for use with displacement-based acceptance criteria shall be the sum of seismic demand, D_S, and non-seismic demand, D_{NS}, as combined with the following load combination:

$$D = D_{NS} + D_S \qquad \text{(Eq. 5-5)}$$

where D, D_{NS}, and D_S are as defined above.

This load combination is used for nonlinear seismic analyses and to evaluate deformations in linear seismic analyses.

5.1.2.3 *Inelastic Energy Absorption Factor,* F_μ

Earthquake demand, given by forces, moments, or stresses (i.e., not deformations) may be reduced by the inelastic energy absorption factor, F_μ, provided in Table 5-1, for the given Limit State. F_μ for Limit State D shall be unity. F_μ for out-of-plane behavior of concrete wall/slab systems, meeting the detailing requirements of Section 6.0, shall be taken from values for concrete moment frames. Linear interpolation shall be used to determine values of F_μ for intermediate values of ℓ/h, f_v, P, and e in Table 5-1. F_μ for compression and shear in columns shall be unity.

SECTION 5.2 ACCEPTANCE CRITERIA

5.2.1 General

(a) Linear analyses shall meet the strength acceptance criteria in Section 5.2.2(a) for the load combination in Eq. (5-1) and the displacement criteria in Section 5.2.3.1(a) for the load combination in Eq. (5-5).
(b) Nonlinear analyses shall meet the displacement acceptance criteria in Sections 5.2.3.1 and 5.2.3.2. The capacity of yielding elements in nonlinear analyses shall meet the strength design criteria in Section 5.2.2(b). The load combination in Eq. (5-5) shall be used in conjunction with the nonlinear acceptance criteria.
(c) For steel structures based on the AISC ASD, the allowable stresses shall be scaled to strength-based nominal code capacity, as described in Section 4.0.
(d) All structures shall meet the ductile detailing requirements of Section 6.0.

5.2.2 Strength Acceptance Criteria

(a) For linear analyses, the total demand acting on an element, D, shall be less than or equal to the element's code capacity:

$$D \leq C \qquad \text{(Eq. 5-6)}$$

where the code capacity is developed in accordance with Section 4.0 and the demand is developed with Eq. (5-1), with seismic demand determined from Section 3.0.

(b) For nonlinear analyses, the capacity of all elements, including yielding elements, shall be limited to code capacities:

ϕM_n for bending,
ϕV_n for shear, and
ϕP_n for axial loads.

M_n, V_n, and P_n are the nominal code capacities in flexure, shear, and axial force, respectively.

5.2.3 Deformation Acceptance Criteria

5.2.3.1 Allowable Drift Limits for Structural Systems

(a) The total story drift ratio for each story shall be less than the allowable values given in Table 5-2 where the total story drift ratio is calculated with the demand in Eq. (5-5) (drift ratio, γ, is the story displacement divided by the story height).

(b) For nonlinear analysis of shear wall or braced frame structures, the total drift ratio of a given element for each story shall be less than the allowable drift ratio:

$$\gamma_t \leq \gamma_a \qquad \text{(Eq. 5-7)}$$

where

γ_t = Total drift ratio of an element (in radians) due to the demand in Eq. (5-5)

γ_a = Allowable drift ratio (in radians) given in Table 5-2

(c) Allowable drift limits for out-of-plane behavior of concrete wall/slab systems shall be taken from values for concrete moment frames.

5.2.3.2 Allowable Rotation Limits for Structural Members.

For nonlinear analysis of frame structures, the nonlinear hinge rotation in beams and columns shall be less than the allowable nonlinear hinge rotation:

$$\theta_N \leq \theta_A \qquad \text{(Eq. 5-8)}$$

where

θ_N = Nonlinear hinge rotation in an element (in radians) due to the demand in Eq. (5-5)

θ_A = Allowable hinge rotation (in radians) given in Table 5-3

Linear interpolation shall be used to determine deformation limits for intermediate values of ℓ/h and of P in Table 5-3.

The allowable hinge rotation for frame members, θ_A, in Table 5-3 represents the nonlinear rotation of a plastic hinge. Columns with $\theta_A = 0$ shall be designed to behave elastically.

SECTION 6.0 DUCTILE DETAILING REQUIREMENTS

The inelastic energy absorption (ductility) factors given in Section 5.0 imply that inelastic deformation is permissible when structures are subjected to extreme loads due to natural phenomena hazards. Inelastic response is limited as a function of the intended Limit State. It should be recognized that inelastic deformation means damage (although limited) to the structure and, after the design basis event, the structure may need to be repaired before it can be placed in full service.

Ductile detailing requirements must be observed in detailing all types of structures and supports for systems and components to ensure acceptable behavior during earthquakes. In addition to requirements covering materials, connections, and anchorages, they

TABLE 5-2. Allowable Drift Limits as a Function of Limit State and Structural Systems

	Allowable Drift Limit, γ_a			
Limit State	LS-A	LS-B	LS-C	LS-D
Reinforced concrete SMRF	0.025	0.015	0.010	0.005
Reinforced concrete shear wall, in plane:				
Bending controlled walls, $\frac{h_w}{\ell_w} \geq 2.0$				
$6\sqrt{f'_c} < f_v$	0.008	0.006	0.004	0.004
$f_v < 3\sqrt{f'_c}$	0.010	0.008	0.005	0.005
Shear controlled walls, $\frac{h_w}{\ell_w} < 2.0$	0.0075	0.006	0.004	0.004
Steel SMRF	0.035	0.025	0.010	0.005
Steel braced frames:				
Concentric	0.020	0.013	0.005	0.005
Eccentric	0.030	0.017	0.005	0.005

TABLE 5-3. Allowable Rotation Limits for Nonlinear Analysis

Limit State		Allowable Nonlinear Hinge Rotation, θ_A*		
		LS-A	LS-B	LS-C
SMRF reinforced concrete moment frames:				
Beams	$(15 \leq \ell/h)$	0.020	0.010	0.005
Beams	$(\ell/h \leq 10)$	0.010	0.0075	0.005
Columns		0.005	0.0025	0.000
SMRF steel moment frames:				
Beams and columns	$P < 0.2\,P_y$	0.030	0.017	0.004
Columns	$P = 0.3\,P_y$	0.021	0.012	0.004
Columns	$P = 0.4\,P_y$	0.013	0.009	0.004
Columns	$P = 0.5\,P_y$	0.006	0.005	0.004
Columns	$P > 0.5\,P_y$	0.000	0.000	0.000
Slab/wall moment frames:				
Roof slabs, floor slabs, beams and walls of reinforced concrete				
	$(15 \leq \ell/h)$	0.010	0.0075	0.005
	$(\ell/h \leq 10)$	0.0075	0.006	0.005

Note: * $\theta_A = 0.00$ for Limit State D.

include the following:

- **Irregularities.** Structure layout and detailing shall be such that both horizontal and vertical irregularities are minimized.
- **Strength distribution.** Members of the lateral load-resisting system shall be designed with nearly uniform design margin for lateral loads. The goal here is to create a system where inelastic deformations are well distributed throughout the structure.
- **Redundancy.** Structures with large redundancy behave better under earthquake loads. Therefore, layout of the structure shall include as much redundancy as practical.
- **Seismic interaction.** SSCs that span between structures shall be designed to accommodate relative seismic motions.

SECTION 6.1 STEEL STRUCTURES

Steel structures are effective in resisting seismic loads as long as they are designed and detailed to behave in a ductile manner. Preferred types of steel structures include moment frames, braced frames, and eccentrically braced frames. In the following subsections, the principles to be followed in detailing each type of steel structure and steel supports for systems and components are described. Use the applicable parts of the AISC Seismic Provisions.

6.1.1 Moment Frames

Moment-resisting steel frames resist the lateral loads in bending. Seismic energy dissipation is through inelastic bending of members at or near the joints. Moment frames are inherently more flexible and thus their design may be controlled by drift limits. Thus, P-Δ effects may impact the behavior, especially for taller structures, and must be accounted for in design.

In order to ensure ductile behavior, the following design and detailing principles shall be followed:

- **Type of moment frame.** For SDC 3, 4, and 5 structures, only special moment-resisting frames shall be used.
- **Strong column–weak beam.** The design shall be based on the strong column–weak beam philosophy as described in ANSI/AISC 341-02. In such a design, seismic strength of columns at a joint is greater than the seismic strength of beams (see ANSI/AISC 341-02). Experience indicates that this design feature precludes catastrophic structural failure.
- **Yielding away from joints.** Yielding at beam ends shall take place away from the joints. Using reduced beam sections (RBS) or strengthening the joints satisfies this condition. FEMA 350 provides guidance on beam–column connection details to ensure ductile behavior.

- **Panel zone shear.** Panel zone shear capacity shall be greater than the flexural capacity of the connected beam(s) so that panel zone shear deformation effects will be small. Guidance in panel zone shear design is provided in FEMA 350 and ANSI/AISC 341-02.
- **Connection design.** Connection design shall ensure adequate strength so that no other element will yield or fail prior to yielding at the beam end. For this purpose, doubler plates and continuity plates shall be provided as needed (see FEMA 350 and ANSI/AISC 341-02).
- **Welding.** When moment frames are designed using welded connections, both the design and construction of the weld must take into account the factors that affect the joint behavior. The North Ridge, California, earthquake showed that many factors might lead to premature failure of a welded joint. These factors include notch toughness of both base and weld material, weld quality, material properties (e.g., yield strength much higher than intended), and bottom backing bar. Guidance for producing good welds, including weld metal toughness specifications, is given in FEMA 350.

6.1.2 Braced Frames

In the case of the braced frames, seismic energy dissipation is through inelastic deformation of the bracing members. Braced frames are more rigid; thus, drift limits seldom control the design. The recommended detailing practices for braced frames include the following:

- **Braced frame types.** Preferred types of steel bracing for SDC 3, 4, and 5 structures are cross bracing, concentric bracing, and eccentric bracing. If chevron bracing is used, it must meet special ANSI/AISC 341-02 seismic provisions. K-bracing shall not be used.
- **Yielding members.** In the case of the braced frame structures, only the bracing members shall be allowed to yield under extreme loads. Columns and beams in the lateral load path shall be designed to remain elastic under seismic loads.
- **Connection details.** Connections shall be detailed such that connection ultimate strength (bolts and welds) can develop greater strength than the bracing yield strength in accordance with the governing material code.
- **Slenderness ratio.** Bracing slenderness ratio shall be limited to $1000/(F_y)^{1/2}$, where F_y is in kips per square inch (ksi).
- **Bracing design.** Bracing members may be designed to act both in tension and compression or in tension only. In the former case, the total tension capacity shall not exceed 70% of the total lateral load on the braced frame. In the latter case, tension bracing shall resist the entire lateral load.
- **Eccentric bracing.** In the design of the eccentrically braced frames, seismic energy dissipation is through the "link" (i.e., shorter segment of the beam that is intended to yield in shear). All other elements in the load path shall be designed for a minimum of 125% of the beam yield strength.
- **Compression members.** Elements of the compression members shall meet the width–thickness ratios of "compact" members as defined by ANSI/AISC 341-02.

SECTION 6.2 REINFORCED CONCRETE

6.2.1 General

Reinforced concrete structures shall be detailed in accordance with the provisions of ACI 349, including Chapter 21, "General Provisions for Seismic Design." Specifically, the following requirements apply to the concrete and reinforcing steel:

- The minimum specified strength for the concrete shall be 3000 psi
- The maximum specified yield strength for reinforcement steel shall be 60,000 psi.
- The yield strength of the reinforcement based on actual mill tests shall not exceed the specified yield strength by more than 18,000 psi.
- The ratio between the reinforcement actual ultimate tensile strength and the actual tensile yield strength shall be 1.25 or more.
- Substitution of reinforcing steel shown on design documents with steel having higher specified yield strength or larger reinforcement areas is not permitted.
- Beams, columns, and walls (out-of-plane forces) that are part of the lateral load-resisting system shall have shear capacity equal to at least that corresponding to 125% of the moment capacity of the member.

6.2.2 Slab/Wall Moment Frame Systems

Slabs and walls in slab/wall moment frames shall meet the following requirements:

- The positive out-of-plane moment strength at the face of the joint shall be not less than one-half the negative moment strength provided at that face of the joint.

- Reinforcement shall develop its yield strength at the face of the support.
- Joints of the slab/wall moment frame systems shall be confined in accordance with one of the following:

(a) Shear and bending moment in the disturbed region of joints shall meet the strut and tie criteria of ACI 318. The strength of diagonal compression struts crossing the joint shall be calculated using $\beta_s = 0.60$. Minimum joint reinforcement shall consist of X-pairs of #4 diagonal crossties spaced 12 in. on center. Diagonal joint crossties shall be anchored with seismic hooks.

(b) Transverse joint reinforcement in walls shall conform to Eq. (11-13) in ACI 349 such that $V_u \leq \phi V_n$, where

$$V_n = 15\sqrt{f'_c}A_j \text{ for interior slab/wall joints}$$

$$V_n = 8\sqrt{f'_c}A_j \text{ for exterior slab/wall joints}$$

(c) Transverse joint reinforcement in walls shall conform to Section 21.4.4 of ACI 349 such that $V_u \leq \phi V_n$, where

$$V_n = 20\sqrt{f'_c}A_j \text{ for interior slab/wall joints}$$

$$V_n = 12\sqrt{f'_c}A_j \text{ for exterior slab/wall joints}$$

where A_j is the effective cross-sectional area within a joint as defined in Chapter 21 of ACI 349.

Beams and columns, which comprise portions of the slab/wall frame, shall be detailed in accordance with Section 21 of ACI 349.

SECTION 6.3 ANCHORAGE

The most reliable anchorage will be achieved by properly installed cast-in-place bolts or headed studs, undercut-type expansion anchors, or welding to embedded plates. Other expansion anchors are not recommended for vibratory environments (i.e., support of rotating machinery), for very heavy equipment, or for sustained tension supports and are prohibited from use for SDC-5. Epoxy grouted anchorage is not allowed in elevated temperature, radiation environments, or overhead applications, and shall not be used without documented qualification following ACI 349. Expansion and epoxy grouted anchorages must be installed following the manufacturer's instructions.

SECTION 7.0 SPECIAL CONSIDERATIONS

SECTION 7.1 ROCKING AND SLIDING OF UNANCHORED RIGID BODIES

It is generally preferable to anchor components so as to prevent rocking and sliding. However, unanchored components are acceptable as long as the provisions of this section are satisfied. When checking sliding, the coefficient of sliding friction shall be set at the 95% exceedance level. When checking rocking, the coefficient of sliding friction shall be set at the 5% exceedance level. Alternately, pure rocking (i.e., no slide–rock) can be assumed.

Median-centered analysis techniques shall be used so as to predict "best-estimate" values of sliding distance and rocking angle. These "best-estimate" values shall be increased by the following Factors of Safety (FS) in order to obtain design values of sliding and rocking:

$$\text{Sliding: } FS_S = 3.0 \qquad \text{[Eq. 7-1(a)]}$$

$$\text{Rocking: } FS_R = 2.0 \qquad \text{[Eq. 7-1(b)]}$$

except that the design value of sliding does not need to exceed 1.5 times the peak displacement of the input motion. Sufficient clearances shall be provided to accommodate the design values of both sliding and rocking. However, these design values of sliding and rocking do not have to be combined. In addition, the design value of rocking angle shall be held less than the instability angle, α, defined by:

$$\alpha = \arctan(a) \qquad \text{[Eq. 7-2(a)]}$$

$$a = \frac{b}{h} \qquad \text{[Eq. 7-2(b)]}$$

where

b = Minimum horizontal distance from the edge of the body to the center of gravity

h = Center of gravity height, as shown in Figure 7-1

"Best-estimate" values of sliding distance or rocking angle may be determined either by time-history analysis or by the approximate methods. An acceptable set of approximate methods is presented in Appendix A of this Standard.

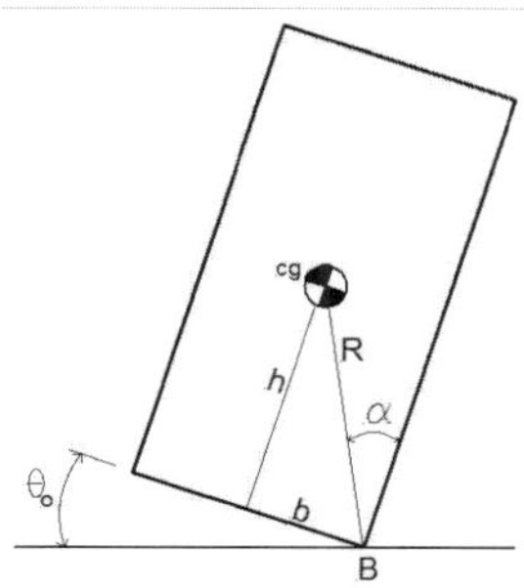

FIGURE 7-1. Rigid Body Rocking Definitions

When time-history analysis is used, a minimum of five different time histories that satisfy the requirements of Section 2.4 shall be used. The mean values of sliding and rocking angle from these multiple time-history analyses shall be used as the "best-estimate" values.

SECTION 7.2 BUILDING SLIDING AND OVERTURNING

7.2.1 Building Sliding

The sliding stability of a building can be checked using a static evaluation, comparing the lateral forces with the static resistance.

$$\text{Static Resistance} > 1.1 \times V_{BaseShear} \quad \text{(Eq. 7-3)}$$

Sliding stability is demonstrated by satisfying this condition. The base shear from linear analysis is determined using the approach given in Section 7.4.1, which accounts for the inelastic energy absorption factor. For foundations, the static resistance is given by

$$V_R = C + N \times \mu + P_u$$

where

C = Effective cohesion force (effective cohesion stress, c, times foundation base area)
N = Normal (compressive) force
μ = Coefficient of friction
P_u = Passive resistance

If included, traction on the side of the foundation is calculated in a similar manner. The coefficient of friction is determined from geotechnical investigations. In the absence of such information, μ may be based on the minimum of the effective internal friction angle of the soil and the friction coefficient between soil and foundation from published references. The passive pressure resistance and the associated deformation shall include internal friction and cohesion, as appropriate.

The horizontal deflection of the foundation is determined using the load-versus-deflection curve for the combined effect of cohesion, friction, and passive resistance. The functionality of the building is checked and, if it is not impaired, an acceptable condition exists.

If the above condition is not satisfied, it is permissible to demonstrate an acceptable condition by estimating the larger foundation movement resulting when the dynamic base shear momentarily exceeds the soil resistance and demonstrating the building remain functional for the resulting displacements. A refined estimate of the foundation movement shall recognize the time-dependent reversing nature of earthquake motion and the nonlinear soil resistance. If the functionality of the building can be demonstrated for the larger displacements, an acceptable condition exists.

The nonlinear sliding analysis method of Section A.1 (in Appendix A of this Standard) can be used to estimate the foundation sliding movement as long as the fundamental frequency, f_b, of the building exceeds the effective sliding frequency, f_{eS}, defined in Section A.1. If not, then substitute f_b for f_{eS} in Eq. (A-3). When using this method, the appropriate coefficient of friction, μ, to be used in Eq. (A-1) is given by

$$\mu = \frac{V_R}{W} \quad \text{(Eq. 7-4)}$$

where

V_R = Total sliding resistance
W = Building weight

7.2.2 Building Overturning

The acceptance criteria for overturning consists of a two-level check, which recognizes that the overturning moment from a fixed-base analysis is conservative. If the restoring moment exceeds the overturning moment, a fixed-base analysis provides a reasonable estimate of the soil pressure. Therefore, building stability is demonstrated if

$$\text{Restoring Moment} > 1.1 \times M_{OT} \quad \text{(Eq. 7-5)}$$

and if the bearing pressure is acceptable. The bearing capacity shall consider settlement criteria to ensure small displacements. The overturning moment from linear analysis is determined using the approach given in Section 7.4.1.

For conditions where the M_{OT} from a fixed-base analysis exceeds the restoring moment, uplift may occur. The uplift condition reduces the contact area and increases the potential for soil instability. The building shall be evaluated for the uplift condition since there is considerable margin beyond Eq. (7-5). The evaluation of the uplift condition shall include the appropriate overturning moment, soil pressure, building movement resulting from the uplift, the effect of the building movement on buried piping, connections between buildings, and other items influenced by the rocking motion. The soil capacity for this evaluation shall be based on ultimate conditions and not limited by settlement criteria. If the associated SSCs remain functional, the overturning condition is acceptable.

The nonlinear rocking analysis method of Section A.2 can be used to demonstrate building stability and for estimating the maximum rocking angle, as long as the effective rocking frequency, f_e, computed from Eq. [A-7(a)] is less than the fundamental frequency of the building soil supported with no uplift. When using this method, the building shall be assumed to rock about a point, B, located at the centroid of the soil contact area, A_C, given by

$$A_C = \frac{W}{q_{ult}} \quad \text{(Eq. 7-6)}$$

where

W = Building weight

q_{ult} = Value based on the ultimate capacity of the soil not limited by settlement considerations.

SECTION 7.3 SEISMIC SEPARATION

A minimum separation of Δ, calculated as follows, shall be maintained between adjacent SSCs:

$$\Delta = 2.0\,(\Delta_1^2 + \Delta_2^2)^{1/2} \quad \text{(Eq. 7-7)}$$

where Δ_1 and Δ_2 are the maximum elastically computed displacements, relative to the base of the foundation of each SSC, along the same axis for the adjacent SSCs.

SECTION 7.4 SEISMIC DESIGN CONSIDERATIONS FOR FOUNDATION ELEMENTS

Foundation elements consist of the supporting soil and structural elements to which the structure is attached. A continuous load path, or paths, through structural members with adequate strength and stiffness shall be provided to transfer all forces from the point of application to the supporting soil and/or pile foundation. The foundation shall be designed to resist the forces developed and accommodate the movements imparted to the structure by the design ground motions. Structural elements of the foundation and their connections should be designed to be ductile. A one-third increase in soil strength above static soil capacity can be used for the structural design of foundation elements subjected to combined seismic and non-seismic forces.

7.4.1 Linear Analyses

The seismic load on the supporting soil and/or pile foundation and connections of the piles and caissons to structural components are the loads from the seismic analysis described in Section 5.1.2.1:

$$D = D_{NS} + D_S$$

The maximum foundation load need not exceed the force that the structure can deliver to it.

7.4.2 Nonlinear Analyses

The seismic distortions on (1) the foundation components and (2) the connections of piles and caissons to structural components are the distortions from the analysis described in Section 3.3. The force-displacement behavior of the foundation elements shall be calculated using the stiffness and strength obtained using established principles of soil mechanics. Variability of stiffness and strength must be considered in evaluating foundation demands to ensure mean estimates of demand on the foundation element.

7.4.3 Special Provisions for Foundation Components

The design and construction of foundation components shall comply with the detailing and design provisions for foundation elements specified for IBC SDC-D.

Piling shall be designed and constructed to withstand maximum imposed curvatures from earthquake ground motions and structure response. Curvatures shall include free field soil strains modified for soil pile structure interaction coupled with pile deformations induced by lateral pile resistance to seismic forces imposed by the structure.

7.4.4 Liquefaction Potential and Soil Strength Loss

The potential consequences of any liquefaction and soil strength loss—including estimation of differential settlement, lateral movement, or reduction in foundation soil bearing capacity—shall be incorporated into the design. Such measures can include, but are not limited to, ground stabilization, selection of appropriate structural systems to accommodate anticipated displacements, consideration of loss of constraint of supporting soil, or any combination of these measures.

The potential for liquefaction and soil strength loss shall be evaluated for site PGAs, magnitudes, and source characteristics consistent with the design earthquake ground motions. The PGA shall be determined based on a site-specific study, taking into account soil amplification effects.

SECTION 7.5 UNREINFORCED MASONRY USED AS MOVABLE PARTITIONS, BARRIERS, AND RADIATION SHIELDING

Unreinforced masonry walls can be found in modern nuclear facilities. The primary uses of these walls are temporary partitions, barriers, and radiological shielding. If such walls are intended to be permanent, or at least not movable, they shall be steel reinforced and anchored. Both in-plane and out-of-plane lateral loads shall be considered in design.

The behavior of free-standing, non–load-bearing unreinforced masonry walls is somewhat different from reinforced masonry or concrete walls in that steel is not available to (1) carry tension induced by out-of-plane bending, (2) bridge across cracks caused by in-plane shear or diagonal tension, and (3) anchor the walls for uplift.

- **In-plane loading.** The requirements of Chapters 5 and 6 of ACI 530 shall be followed for the design of unreinforced masonry walls. However, ACI 530 does not provide any criteria for design of unreinforced masonry walls subject to in-plane loads where the aspect ratio is such that in-plane overturning moments result in tensile uplift loads on the wall. To ensure no uplift, the height-to-length ratio shall be less than 0.25. When this value is exceeded, positive tensile anchorage shall be provided to vertically anchor the wall. The shear or diagonal tension in-plane load-carrying capacity shall be as given in Section 6.5.2 of ACI 530, multiplied by 1.6.
- **Out-of-plane loading.** For out-of-plane loading on the unreinforced wall, the resultant tensile loads shall be limited to the values shown in Table 6.3.1.1 of ACI 530 multiplied by 1.6.

SECTION 7.6 PROVISIONS FOR CONSTRUCTION EFFECTS

Large and heavy nuclear structures with foundation basemat and/or sidewalls in contact with soil and backfill materials may be subjected to substantial additional bending effects due to differential settlements caused by the heavy loads, soft or nonuniform site conditions, site excavation effects, anticipated groundwater variation during construction, or construction sequence effects. If geotechnical evaluation indicates that the induced settlements can be large (greater than several inches across the site), the potential effect of such settlements on bending in the foundation elements needs to be evaluated. This should include an evaluation of construction sequence (concrete pour sequence) to limit such potential differential settlement effects. Reinforcement design and detailing shall consider the potential effects of such differential settlements and minimize the potential for induced cracking of the foundation elements (walls and basemat).

Embedded walls shall be designed to take into account anticipated soil pressures expected during construction and normal operating and seismic conditions. Under normal operating conditions, lateral wall pressures are typically estimated as the sum of (a) at-rest soil pressures, (b) effects of compaction of adjacent soils (if any), and (c) pressures induced by adjacent structures. Dynamic pressures from seismic effects can be evaluated following the elastic procedures described in ASCE 4. Additional dynamic pressures from adjacent structures, if any, need to be incorporated. If passive pressure assumptions are used in the evaluation of potential sliding of the embedded structure, the walls must also be evaluated to ensure that they can sustain such pressures.

SECTION 8.0 EQUIPMENT AND DISTRIBUTION SYSTEMS

SECTION 8.1 INTRODUCTION

Seismic qualification shall be based on equivalent-static or dynamic analysis, testing, past earthquake experience, or generic test data. Permanent equipment and distribution systems shall be adequately supported and anchored. Whenever any one or a combination of these methods is chosen, it must be demonstrated that the equipment or distribution systems will be capable of performing all of their specified safety functions. It shall also be verified that the equipment and distribution systems do not interfere with the safety function of adjacent equipment or distribution systems per ANS 2.26. The seismic demand for equipment and distribution systems is the seismic in-structure response spectra or time history per Section 3.0 of this Standard.

SECTION 8.2 QUALIFICATION BY ANALYSIS

8.2.1 Seismic Analysis Methods

Seismic qualification by analysis can be either an equivalent-static analysis or a dynamic analysis. The following provides the minimum requirements for each approach. The development of seismic demand shall satisfy Section 3.0 of this Standard.

8.2.1.1 Equivalent-Static Analysis

In this type of analysis, seismic inertia loads are statically applied as external forces to a static representation of the equipment or distribution system including supports and anchorage in order to determine resultant primary (equilibrium) stresses and displacements. Equivalent-static loads shall be developed from the peak of the in-structure response spectra at the location of the item considered and appropriate for the SDC of that equipment or distribution system. Analytical procedures for equipment and distribution systems shall meet the requirements of Section 3.2.5, "Equivalent-Static Method," of ASCE 4. In addition, Section 3.2.1 of this Standard shall be met for single-point-of-attachment cantilever components. For equivalent-static analysis of piping systems, paragraph N-1225, "Simplified Dynamic Analysis," of the ASME Boiler and Pressure Vessel Code (B&PVC), Section III, Division 1, Appendix N, shall be used.

Support displacement of multiple supported equipment or distribution systems due to earthquake motion, termed Seismic Anchor Motions (SAM), shall be considered as required in Section 3.2.6 of ASCE 4 for equipment and distribution systems and in paragraph N-1225 of ASME B&PVC, Section III, Division 1, Appendix N, for piping systems.

8.2.1.2 Dynamic Analysis

For dynamic analysis, seismic loads and resultant inertia and seismic anchor motion stresses in equipment and distribution systems shall be based on the earthquake dynamic input (i.e., in-structure response spectra or time histories) appropriate for the SDC for the equipment or distribution system. In addition, the following specified dynamic analysis procedures shall be used as appropriate for the type of component:

- Appendix N of Section III of the ASME B&PVC
- ASCE 4
- IEEE 344
- ASME QME-1

8.2.2 Demand for Qualification by Analysis

The following subsections provide the requirements to define demand on equipment and distribution systems when using qualifications by analysis.

8.2.2.1 Damping

Damping values specified in Table 3-2 shall be used. Damping as percent of critical is defined for the type of component and the Response Level, which is defined in Table 3-4.

8.2.2.2 Inelastic Energy Absorption Factor, F_μ

The inelastic energy absorption factors for equipment and distribution systems allowed for ductile-type behavior under seismic loads is provided in Table 8-1. If the component contains brittle material in the load path or brittle material is used that could affect its specified safety function, then F_μ values shall be taken as 1.0. When seismic analysis methods are used to qualify active components (i.e., components that must change state as part of their safety function) during the earthquake, only Limit State D is permitted. Limit States D or C are permitted for active components if their change in state occurs following the earthquake to include a period sufficient to determine if there is any damage to the plant or as necessary to reset tripped devices, such as relays. Pressure-retaining mechanical equipment that must retain its leak-tight integrity is not permitted in Limit State A.

8.2.2.3 Total Demand for Qualification by Analysis

The total demand on equipment and distribution systems qualified by analysis is given by the following equation:

$$D = D_{NS} + \frac{D_S}{F_\mu} \qquad \text{(Eq. 8-1)}$$

where

D = Total demand acting on the component.

D_{NS} = Non-seismic demand acting in the component (e.g., dead weight and fluid pressure). D_{NS} shall be consistent with the load combinations de-

TABLE 8-1. Equipment and Distribution Systems Inelastic Energy Absorption Factor, F_μ[1]

	Factor, F_μ		
	Limit State A[2]	Limit State B[2]	Limit State C[2]
Equipment:			
Vessel	1.50	1.25	1.15
Heat exchanger	1.50	1.25	1.15
Coolers	1.50	1.25	1.15
Chillers	1.50	1.25	1.15
Tanks (vertical)	1.25	1.25	1.15
Tanks (horizontal)	1.50	1.25	1.15
Pumps	1.50	1.25	1.15
Fans	1.50	1.25	1.15
Valves	1.50	1.25	1.15
Dampers	1.50	1.25	1.15
Filters[3]	2.00	1.50	1.25
Glove boxes[3]	2.00	1.50	1.15
Electrical boards[3]	2.00	1.50	1.15
Electrical racks[3]	2.00	1.50	1.15
Electrical cabinets[3]	2.00	1.50	1.15
Distribution systems:			
Butt joined groove welded pipe	1.75	1.50	1.25
Socket welded pipe	1.50	1.25	1.15
Threaded pipe	1.25	1.15	1.00
Conduit	1.50	1.35	1.25
Instrument tubing	1.50	1.35	1.25
Cable trays	1.50	1.35	1.25
HVAC duct	1.50	1.25	1.15
Equipment supports[3]	2.00	1.50	1.25

Notes:

[1] These inelastic energy absorption factors, $F\mu$, are applicable to equipment functioning in a passive mode. For active components, the Limit State is restricted to Limit State D, where $F\mu = 1$. See Section 8.2.2.2.

[2] Except as discussed in note 3, below, the allowable behavior limits for passive components are based on the ASME B&PVC, Section III, allowables for Service Level D. It should be noted that ASME B&PVC, Section III, Service Level D allowable stresses range from 1.6 to 2.0 σ_y.

[3] These components are normally designed to AISC allowables, which are typically limited to 0.8 to 1.0 σ_y; hence, they are allowed a somewhat higher inelastic energy absorption factor as compared to ASME B&PVC allowables, where allowable stresses can be as high as 2.0 σ_y.

fined in the industry standards applicable to the capacity definition of Section 8.2.3.1, below.

D_S = Seismic response of the component due to the seismic inertia loads from an equivalent-static analysis or from a dynamic analysis, either of which is generated on input from DBE based on the SDC of the component using an elastic analysis of the supporting structure, with F_μ of the supporting structure set to 1.0. The calculated seismic response of the component, D_S, shall be based on its effective natural frequency, f_e, as defined in Eq. (5-4) and on its damping value per Section 8.2.2.1.

When displacements need to be calculated as part of the seismic qualification by analysis, the demand given in Eq. (8-1) shall be changed by replacing D_S/F_μ with D_S.

8.2.3 Capacity Using Qualification by Analysis

The following subsections define the capacity, *C*, of equipment and distribution systems when seismic qualification is based on analysis methods.

8.2.3.1 Capacity Defined by Industry Standards

Capacity shall be based on the specified stress or load limits specified in the appropriate code or standard for the SDC 3, 4, and 5 equipment or distribution systems being qualified. If the capacity limits of ASME nuclear codes or standards (B and PVC, QME-1) are used to define capacity, then all ASME nuclear code construction requirements shall be met except nuclear code stamping.

8.2.3.2 Stress Acceptance Criteria for Capacity

Where ASME B&PVC, Section III, is specified, the ASME B&PVC, Section III, Service Level D is allowed for Limit States A and B for passive components only. For active components, the ASME B&PVC, Section III, Service Level shall be justified to ensure, with high confidence, the performance of the required safety functions. Limiting the calculated stresses to be less than the yield stress of the material will satisfy this requirement for active components. Stress limits shall be limited to be less than yield in components, supports, and anchorage for active components whose design is governed by AISC (see Section 4.2.4).

8.2.4 Acceptance Criteria and Documentation for Qualification by Analysis

Acceptance criteria for qualification by analysis shall clearly demonstrate that the capacity, *C*, exceeds the total demand, *D*, as follows:

$$D \leq C \qquad \text{(Eq. 8-2)}$$

Documentation shall meet the requirements of the specific code or standard, and as a minimum the documentation requirements for analysis as specified in IEEE 344 shall be met.

SECTION 8.3 QUALIFICATION BY TESTING AND EXPERIENCE DATA

8.3.1 Tests and Experience Methods

The following subsections provide the minimum general requirements, acceptance criteria, and governing references for qualification by testing and by use of experience data.

8.3.1.1 Testing

Seismic qualification by testing can be either a proof test or a fragility test. A seismic proof test is performed to determine whether the seismic test input, usually defined as a Test Response Spectrum (TRS), exceeds the specific required seismic input, usually defined as a Required Response Spectrum (RRS). Seismic fragility testing is usually performed in steps of increased levels of seismic input until a failure or malfunction occurs or until the test limits of the test table are reached. In any case, the test specimen shall be mounted on the test table (i.e., shake table) as close as possible to the mounting in the nuclear facility. Any other non-seismic loads that exist concurrently with the seismic loading and that could affect its safety function shall be accounted for in the test program. Seismic qualification by testing shall meet the requirements of IEEE 344 or ASME QME-1, depending on the type of item being qualified. In harsh environments, where the equipment safety function can be affected, seismic testing is only one portion of the qualification program as defined in IEEE 323. Testing of in-line components, such as valve actuators, shall meet IEEE 382.

8.3.1.2 Test Experience Data

Available seismic test data on a specific class of equipment can be gathered, evaluated, and consolidated. These test data can be reduced to a Test Experience Spectrum (TES), which defines the seismic response acceleration levels below which the class of equipment can be expected to function. The TES is acceptable if all the specific equipment class caveats and inclusion rules are met and if it is documented that there are no design or material changes that can reduce the seismic ruggedness of the candidate equipment item when compared to those items tested that form the TES. The application of test experience methodology for seismic qualification in IEEE 344 is acceptable.

TES for chatter-sensitive devices (e.g., relays) for function during an earthquake shall not be used due to the uncertainty associated with vintage issues.

8.3.1.3 Earthquake Experience Data

Seismic experience data concerning the response of mechanical and electrical equipment and distribution systems (e.g., raceway systems) to strong motion seismic excitations can be collected and evaluated and form the basis of the use of earthquake experience data as a seismic qualification method. The procedures associated with verifying seismic adequacy using earthquake experience as given in ASME QME-1 (Section QR and Appendix QR-A) are acceptable. Application of the earthquake experience methodology for seismic qualification specified in IEEE 344 is also acceptable. These procedures specify caveats and inclusion rules that must be satisfied.

The application of earthquake experience, including operability, does not eliminate or substitute for leak-tight and integrity requirements of ASME B&PVC or B31 codes.

Items of newer vintage must be investigated to ensure they have no features that would reduce their seismic ruggedness as defined by the seismic experience data. If the candidate equipment item has complex features and its design varies significantly with time, then other methods of seismic qualification shall be used.

8.3.2 Demand for Qualification by Tests and Experience Data

8.3.2.1 Demand for Qualification by Test and Test Experience Data

The demand for qualification by test is the following:

$$D = D_{NS} + 1.4\,D_S \qquad \text{(Eq. 8-3)}$$

where

D = Total demand on the component.

D_{NS} = Any additional loads that are required during the test (e.g., pressure or voltage). Also, D_{NS} could be any pre-aging required before the seismic test. Typically, $D_{NS} = 0$, other than the dead weight of components.

D_S = Seismic demand from an elastic seismic analysis of the supporting structure per Section 3.0 for the specified SDC for the component being qualified. The support structure $F\mu$ is set to 1.0. Typically, D_S is defined as the in-structure response spectra of the support location of the component.

The factor 1.4 is the equipment capacity factor for qualification by test or TES that provides the margin to obtain the required confidence level of performance.

There is no inelastic energy absorption factor, $F\mu$, since the Limit State is indirectly specified for the component (e.g., functional requirements or structural integrity) that defines its acceptance criteria.

8.3.2.2 Demand for Qualification by Earthquake Experience Data

The demand for qualification by use of earthquake experience data is given by the following equation:

$$D = D_{NS} + D_S \qquad \text{(Eq. 8-4)}$$

D and D_S are defined in Section 8.3.2.1. The implied factor of 1.0 on D_S is the equipment capacity factor for qualification by use of earthquake experience data. Earthquake experience data addresses equipment earthquake performance under normal operating conditions as seen in industrial facilities. Therefore, D_{NS} is defined as normal operating loads and is explicitly incorporated in the earthquake performance of the equipment; for that reason, it is set to zero in Eq. (8-4). If the candidate equipment item has unique non-seismic loads not typical of normal operating conditions, the earthquake experience method of qualification cannot be used without justification that the nontypical load condition will not affect seismic ruggedness or function.

8.3.3 Capacity Defined for Seismic Qualification by Tests and Experience Data

Capacity, C, by testing is defined as the applied test motion at its support location for which the component demonstrated its required function. This is typically defined as the TRS. The specific requirements concerning test capacity shall satisfy the latest revision of IEEE 344 or ASME QME-1, as appropriate for the component being qualified.

Capacity, C, based on test experience data is defined by the TES developed by the equipment seismic test data for a specific and limited set of equipment that was tested. The TES is acceptable provided all the caveats and inclusion rules are met and design changes (vintage) are appropriately evaluated to show no reduction in seismic ruggedness.

Capacity, C, based on earthquake experience data is defined as the Earthquake Experience Spectrum (EES), which is a definition of the equipment capacity based on earthquake ground motion at the database sites used to develop an equipment class. The EES is acceptable when the component meets the equipment class caveats and inclusion rules and design changes (vintage) are appropriately evaluated to show no reduction in seismic ruggedness.

8.3.4 Acceptance Criteria and Documentation for Qualification by Tests and Experience Data

Acceptance criteria for qualification by test or experience data are defined by Eq. (8-2). Acceptability of capacity not exceeding demand at a particular frequency or range of frequencies shall satisfy IEEE 344 and ASME QME-1, as appropriate to the type of component or method of qualification. The specific functionality of the component must be demonstrated by the qualification method used.

Documentation shall meet the requirements of the standard, code, or procedure used. The documentation shall clearly document that capacity is greater than demand for the specified functional requirements.

SECTION 9.0 SEISMIC QUALITY PROVISIONS

The seismic analysis and design of nuclear facilities specified in this Standard will be performed under the purview of the U.S. Department of Energy (DOE) or the U.S. Nuclear Regulatory Commission (NRC). The DOE and NRC have regulatory quality assurance (QA) requirements that are applicable throughout the planning and performance of design activities including seismic analysis. Verification of both design and construction is essential to ensure acceptable seismic behavior, and thus it is an integral part of the QA process.

SECTION 9.1 DESIGN VERIFICATION AND INDEPENDENT PEER REVIEW

9.1.1 Seismic Design Verification

Design verification shall include provisions for verifying and checking the adequacy of the analysis and design by any one or a combination of the following:

1. Using alternate or simplified calculation methods,
2. Using a suitable testing program, or
3. Checking the original analysis and verifying the underlying assumptions.

Individuals or groups other than those who performed the design work must perform design verification. The ANS/ASME NQA-1, Quality Assurance Requirements for Nuclear Facility Applications, includes detailed design verification processes that should be used for complying with the QA portion of this Standard. DOE regulatory guidance from G 414.1-2, Quality Assurance Management System Guide, is also appropriate for this purpose. NRC QA criteria of Appendix B to 10 CFR Part 50 also provides guidance on software verification.

9.1.2 Independent Seismic Peer Review

An independent seismic peer review shall be conducted for SDC 3, 4, and 5 SSCs. Independent seismic peer review is a form of design review and shall be conducted by qualified individuals with knowledge of the unique features of the design and analysis as well as the type of construction employed. The peer review program shall be developed using a graded approach, paralleling the seismic design categories of 3, 4, and 5, so the level of review is commensurate with the complexity of the design, the number of design disciplines involved, and any innovative analysis and design concepts employed. The peer reviewers shall be independent from the design verifiers.

SECTION 9.2 STRUCTURAL OBSERVATION, INSPECTION, AND TESTING

Construction of a nuclear facility in accordance with the design documents is essential to ensure that SSCs respond to an earthquake as intended. A verification program shall be implemented that is consistent with the applicable provisions of the design code of record and that is based on a graded approach that will address the elements described in the following subsections.

9.2.1 Structural Observations

The design organization shall perform structural observations for critical construction processes. The stages of structural observations (defined as observation of the structural system by the design organization at significant construction stages for conformance to the approved plans and specifications), their scope, and the type of records to be kept shall be identified in the verification program. Records of the observation and resolution of any comments shall be entered into the permanent project records.

9.2.2 Continuous and Periodic Inspections

Personnel who are qualified in accordance with requirements identified by the design organization shall perform inspections. Duties include performing visual inspections and field measurements of materials, obtaining specimens for testing, and preparing reports. The verification program shall identify the items requiring both continuous and periodic inspections. It should also include a schedule of inspections and should specify the records to be kept and reports to be filed. The inspection program shall include foundation and backfill soils, reinforced concrete, structural masonry, and steel structures as well as anchorage and supports that are part of these structures. It should also identify permissible equipment to be used and standards to be followed for processes subject to inspection. Items requiring continuous and periodic examination shall be clearly identified, including the frequency of inspection.

9.2.3 Testing

The verification program shall identify in-progress testing requirements and methods of testing. All testing shall be performed by testing laboratories under the responsible charge of a licensed professional engineer, approved by the design organization. Testing shall include measuring, examining, testing, calibrating, or otherwise determining the characteristics or performance of construction materials and verifying compliance with construction documents. The verification program shall also specify the number and frequency of taking of test specimens in accordance with the applicable codes and standards or as specified in the approved plans, whichever is more stringent.

Records of inspection and testing shall be reviewed by the design organization and made a part of the permanent records for the facility.

SECTION 9.3 QUALITY ASSURANCE

A QA program that, at a minimum, achieves the DOE or NRC QA requirements for organization, training and qualifications, work processes, documents and records (e.g., specifications, drawings, procedures, and instructions), and assessments shall be implemented by projects designed in accordance with this Standard. The QA program shall include the minimum required elements described in the following sections.

9.3.1 Design Basis Documents

Design basis documents shall provide details on the following:

- Design organization, including design team, division of responsibility, team interface control, organizational procedures and standards, design basis documents, instructions, procedures, and drawings
- Design input and specification (criteria documents), including design basis loads, interfaces, load combinations, and acceptable structural performance (i.e., acceptable ductility for structures and structural subsystems)
- Industry standards and codes applicable to the design
- Analysis assumptions and methodology for structures and structural subsystems
- Design report to the owner
- Design output documents

9.3.2 Design Procedures

Procedures are required that provide for the following:

- Verification of analytical software
- Control and updating of software
- Design approval and report
- Design change and document control
- Design audits/independent assessments
- Design verification methods
- Independent peer review

A graded approach shall be used in establishing the QA program and it shall be performed with increasing rigor from SDC-3 to SDC-5.

APPENDICES

APPENDIX A.0 APPROXIMATE METHODS FOR SLIDING AND ROCKING OF AN UNANCHORED RIGID BODY

Commentary and examples on approximate methods for sliding and rocking of an unanchored rigid body can be found in Appendix B.

SECTION A.1 APPROXIMATE METHOD FOR SLIDING OF AN UNANCHORED RIGID BODY

A conservatively biased sliding distance estimate for a rigid body sliding on a level surface may be computed by the following procedure. First, define an effective coefficient of friction, μ_e, by:

$$\mu_e = \mu\,[1 - 0.4\,A_V/g] \qquad \text{(Eq. A-1)}$$

where μ is the coefficient of sliding friction and A_V is the peak vertical acceleration. Next, define a sliding coefficient, c_S, by:

$$c_S = 2\mu_e\, g \qquad \text{(Eq. A-2)}$$

where g is the acceleration due to gravity. The conservatively biased "best estimate" sliding distance, δ_S, is given by:

$$\delta_S = \frac{c_S}{(2\pi f_{eS})^2} \qquad \text{(Eq. A-3)}$$

where f_{eS} is the lowest natural frequency at which the horizontal 10% damped vector spectral acceleration SA_{VH} equals c_S, where

$$SA_{VH} = \left[SA_{H_1}^2 + 0.16SA_{H_2}^2\right]^{1/2} \qquad \text{(Eq. A-4)}$$

in which SA_{H_1} and SA_{H_2} are the 10% damped spectral accelerations for each of the two orthogonal horizontal components, where SA_{H_1} is the larger of the two spectral accelerations.

When this conservatively biased approximate method is used to estimate sliding displacements, the required sliding factor of safety, FS_S, may be reduced to 2.0 because of the conservatism of this approximate method.

SECTION A.2 APPROXIMATE METHOD FOR ROCKING OF AN UNANCHORED RIGID BODY

The "best estimate" maximum rocking angle, θ_o (see Figure A-1), may be computed by the following procedure. In the following procedure, θ_o is always taken as positive (i.e., the absolute value of the maximum rocking angle is used). In some situations, the lateral inertial mass, M_L, inducing lateral rocking forces, differs from the vertical mass, M, resisting rocking. The following approach accommodates the possible difference between M_L and M.

Step #1 Determine the horizontal spectral acceleration capacity, SAH_{CAP}, corresponding to any rotation angle, θ_o.

$$SAH_{CAP} = \frac{2g(f_1(\theta_o) - 1)}{F_H F_V \theta_o} \qquad \text{(Eq. A-5)}$$

$$f_1(\theta_o) = \cos(\theta_o) + (a)\sin(\theta_o) \qquad \text{[Eq. A-5(a)]}$$

$$F_H = \frac{M_L h_L}{Mh} \qquad \text{[Eq. A-5(b)]}$$

$$F_V = \left[1 + \left(\frac{a(SAV)}{F_H(SAH)}\right)^2\right]^{1/2} \qquad \text{[Eq. A-5(c)]}$$

where

g = Acceleration due to gravity
a = As defined in Eq. [7-2(b)]
F_V = Correction for probabilistically combined vertical ground motion
SAV, SAH = Ratio of vertical to horizontal spectral acceleration determined at the effective rocking frequency, f_e, and effective damping

The factor F_H corrects for the difference between the lateral inertial mass, M_L, and the vertical resisting mass, M, where h_L is the center of gravity height for the lateral inertial mass, M_L, and h is the center of gravity height of the resisting mass, M.

For θ_o angles less than 0.4 rad, the following approximation can be used:

$$(f_1(\theta_o) - 1) \approx \theta_o\left(a - \frac{\theta_o}{2}\right) \qquad \text{[Eq. A-5(d)]}$$

$$SAH_{CAP} \approx \frac{g}{F_H F_V}(2a - \theta_o) \qquad \text{[Eq. A-5(e)]}$$

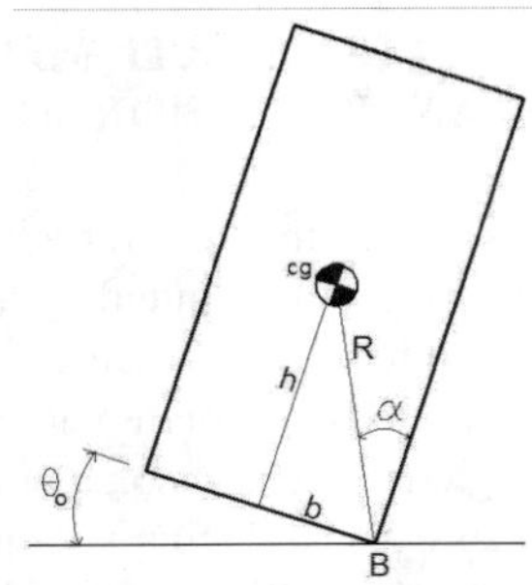

A-1. Rigid Body Rocking Definitions

Step #2 Determine the effective rocking frequency, f_e, corresponding to any θ_o and effective damping, β_e.

$$f_e = \frac{1}{2\pi}\left[\frac{2(f_1(\theta_o) - 1)g}{C_I\theta_o^2 h}\right]^{1/2} \qquad \text{[Eq. A-6(a)]}$$

$$\beta_e = \frac{\gamma}{[4\pi^2 + \gamma^2]^{1/2}} \qquad \text{[Eq. A-6(b)]}$$

where

$$C_I = \left(\frac{I_B}{Mh^2}\right) \qquad \text{[Eq. A-6(c)]}$$

$$\gamma = -2\ell n(C_R) \qquad \text{[Eq. A-6(d)]}$$

$$C_R = \left[1 - \frac{2a^2}{C_I}\right] \qquad \text{[Eq. A-6(e)]}$$

in which I_B is the mass moment of inertia of the rigid body about the edge, B, or center of rotation (see Figure A-1), and M is the rigid body vertical resisting mass. For the situation where the center of gravity is at the center of the rigid body and the lateral inertial mass, M_L, and vertical resisting mass, M, are equal and uniformly distributed,

$$C_I = \frac{4}{3}(1 + a^2) \qquad \text{[Eq. A-6(f)]}$$

Step #3 Determine θ_o for which SAH_{CAP} equals the input spectral acceleration demand, SAH_{DEM}, determined at frequency, f_e, and damping, β_e.

The procedure is to initially determine the frequency, f_{em}, at which the input spectral acceleration demand, SAH_{DEM}, is maximum and back compute θ_{om} from f_{em} using Eq. [A-6(a)]. Next, compute SAH_{CAP} for θ_{om} from Eq. (A-5). If SAH_{CAP} exceeds the maximum SAH_{DEM}, rocking will be negligible. If SAH_{CAP} is less than the maximum SAH_{DEM}, then gradually increase θ_o until SAH_{CAP} first equals SAH_{DEM}; that is,

$$SAH_{CAP} = SAH_{DEM} \qquad \text{(Eq. A-7)}$$

Here, SAH_{DEM} is no longer the maximum input spectral acceleration but is the spectral acceleration evaluated at β_e and f_e, which is a function of θ_o. The resulting θ_o represents the "best estimate" of the maximum rocking angle.

APPENDIX B.0 COMMENTARY ON AND EXAMPLES OF APPROXIMATE METHODS FOR SLIDING AND ROCKING OF AN UNANCHORED RIGID BODY

SECTION B.1 APPROXIMATE METHOD FOR SLIDING OF AN UNANCHORED RIGID BODY

Figure B-1 shows the resisting force, F_{RS}, displacement diagram for a rigid body of mass, M, with sliding resisted by an effective friction coefficient, μ_e, where

$$F_{RS} = \mu_e Mg \qquad \text{(Eq. B-1)}$$

Also shown in Figure B-1 is an equivalent linear force-deflection stiffness, K_e, which absorbs the same work done when displaced, δ_S, where

$$K_e = \frac{2F_{RS}}{\delta_S} = \frac{2\mu_e Mg}{\delta_S} = \frac{c_S M}{\delta_S} \qquad \text{(Eq. B-2)}$$

where c_S is a sliding coefficient defined in Eq. (A–2).

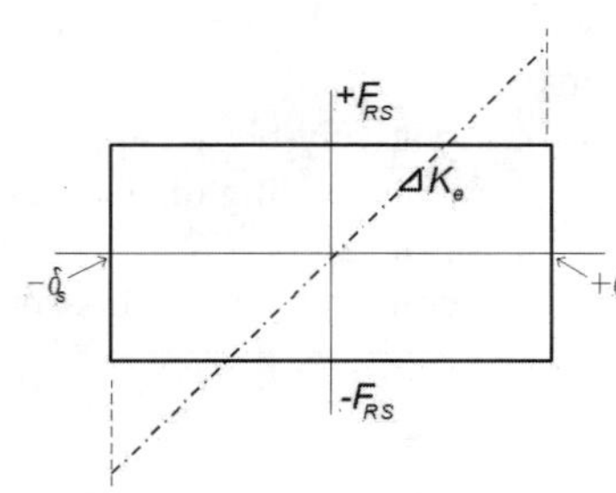

B-1. Sliding Force-Displacement Diagram

The effective frequency of this equivalent linear system is

$$f_{eS} = \frac{1}{2\pi}\left[\frac{K_e}{M}\right]^{1/2} = \frac{1}{2\pi}\left[\frac{c_S}{\delta_S}\right]^{1/2} \qquad \text{(Eq. B-3)}$$

and the vector horizontal spectral acceleration, SA_{VH}, which would displace this equivalent linear system a distance, δ_S, is

$$SA_{VH} = \frac{K_e \delta_S}{M} = c_S \qquad \text{(Eq. B-4)}$$

Thus, for this equivalent linear system,

$$\delta_S = \frac{c_S}{(2\pi f_{eS})^2} \qquad \text{(Eq. B-5)}$$

where f_{eS} is the *lowest* natural frequency at which SA_{VH} equals c_S.

For a complete cycle from $+\delta_S$ to $-\delta_S$ to $+\delta_S$, the equivalent viscous damping, β_H, required to dissipate the total hysteretic energy is

$$\beta_H = \frac{1}{\pi} \approx 0.32 \qquad \text{(Eq. B-6)}$$

or 32%. However, each sliding cycle does not displace the full amount from $+\delta_S$ to $-\delta_S$ to $+\delta_S$ so that the effective damping, β_e, during a time history of response will be less than β_H. Based on a large number of time-history analyses, Ref. [B-1] has suggested the following:

$$\beta_e \approx 0.3\beta_H \approx 10\% \qquad \text{(Eq. B-7)}$$

Lastly, reducing μ_e by $0.4A_V$ in Eq. (A-1) is likely to be excessively conservative for a "best-estimate" sliding displacement. This is true particularly when f_{eS} is low, since the vertical acceleration will oscillate several times during the time the rigid body displaces from zero to δ_S. It is likely that no correction of μ should be made. However, the likely conservatism of Eq. (A-1) will be retained at this time.

Comparison of Sliding Displacement Approaches

Sliding displacements computed by the Reserve Energy approach outlined in Section A.1 versus two Newmark approaches, described in Refs. [B-7] and [B-8], are compared here. The comparison is made for a broad-frequency ground motion response spectrum since the Newmark approaches are only applicable for such an input motion. Only one horizontal direction motion is considered in the comparison, which is consistent with the Newmark approaches.

The input ground motion is assumed to have the following properties:

- PGA: $A = 1.0$ g $= 386$ in./s^2
- PGV: $V = 36$ in./s
- PGD: $D = 20$ in.

and the 10% damped response spectral accelerations, SA, are assumed to be defined by the following:

$$SA = 633 \text{ in./s}^2 \qquad 2.04 \text{ Hz} \le f_{eS} \le 8 \text{ Hz}$$

$$SA = (310 \text{ in./s}) f_{eS} \qquad 0.327 \text{ Hz} \le f_{eS} < 2.04 \text{ Hz}$$

$$SA = (947.5 \text{ in.}) f_{eS}^2 \qquad f_{eS} < 0.327 \text{ Hz}$$

Comparisons are made for effective sliding coefficients of friction, μ_e, of 0.2, 0.4, and 0.7. The Newmark I equation for sliding displacement, δ_S, is

$$\delta_S = \frac{V^2}{2g\mu_e}\left[1 - \frac{\mu_e g}{A}\right] \qquad \text{(Eq. B-8)}$$

and the results are shown in Table B-1. The Newmark II equation for sliding displacement is

$$\delta_S = \frac{2V^2}{g\mu_e}\left[1 - \frac{\mu_e g}{A}\right]^2 \qquad \text{(Eq. B-9)}$$

and the results are also shown in Table B-1. The Reserve Energy approach is described in Section A.1. Results are computed in Table B-2 and also are shown in Table B-1.

The Reserve Energy approach for computing sliding displacements is conservatively biased, as opposed to being a "best estimate" approach. Therefore, when this approach is used, the sliding factor of safety, FS_S, could be reduced to 2.0. Within the large scatter of time-history results, both Newmark II and Reserve

TABLE B-1. Comparison of Sliding Displacements, δ_S

	δ_S (in.)		
μ_e	Newmark I	Newmark II	Reserve Energy
0.2	6.72	21.49	15.77
0.4	2.52	6.04	7.88
0.7	0.72	0.86	4.50

TABLE B-2. Computation of Sliding Displacements, δ_S, by Reserve Energy Approach

μ_e	c_S (in./s^2)	f_{eS} (Hz)	δ_S (in.)
0.2	154.4	0.498	15.77
0.4	308.8	0.996	7.88
0.7	540.4	1.743	4.50

Energy give similar results for $(\mu_e g/A)$ ratios between 0.2 and 0.5 so that this reduced $FS_S = 2.0$ should also be applicable to Newmark II within this range. Newmark I is unconservative for real time histories because real time histories tend to ratchet sliding in one direction. Therefore, a $FS_S = 3.0$ may not be sufficiently conservative when Newmark I is used. Use of Newmark I is not recommended because of this unconservatism.

Since time-history computed sliding displacements are so highly variable, it is impossible to judge between Newmark II and Reserve Energy. This Standard follows the Reserve Energy approach, despite the fact that it is conservatively biased. It has the advantage of being able to be used with floor spectra input.

SECTION B.2 APPROXIMATE METHOD FOR ROCKING OF AN UNANCHORED RIGID BODY

The rocking equation of motion for rotation about point B in Figure A-1 is

$$I_B\ddot{\theta} + M(g + \ddot{y})R\sin(\alpha - \theta) = -MRF_H\cos(\alpha - \theta)\ddot{x} \quad \text{(Eq. B-10)}$$

where

$\ddot{\theta}$ = Rotational acceleration
θ = Rotational angle
$\ddot{x}$ = Horizontal input acceleration
$\ddot{y}$ = Vertical input acceleration
I_B = Mass moment of inertia about point B of the rocking body
M = Mass of the body
α = Given by Eq. [7-2(a)]
F_H = Given by Eq. [A-5(b)]
R = Inclined length, given by

$$R = [b^2 + h^2]^{1/2} = h[1 + a^2]^{1/2} \quad \text{(Eq. B-11)}$$

where a is given by Eq. [7-2(b)].

For positive angle θ, Eq. (B-1) may be rewritten as

$$C_I\ddot{\theta} + \frac{g}{h}f_2(\theta) = -f_1(\theta)F_H\frac{\ddot{x}}{h} - f_2(\theta)\frac{\ddot{y}}{h} \quad \text{(Eq. B-12)}$$

where C_I is defined by Eq. [A-6(c)] and $f_1(\theta)$ and $f_2(\theta)$ are defined by

$$f_1(\theta) = \cos\theta + a\sin\theta \quad \text{[Eq. B-13(a)]}$$

$$f_2(\theta) = a\cos\theta - \sin\theta \quad \text{[Eq. B-13(b)]}$$

Eq. (B-12) is a nonlinear equation of motion. However, by the Reserve Energy method, the left-hand side of this nonlinear equation is approximated by a linear equation that retains the same potential energy at maximum rotation, θ_o, as does the nonlinear equation. The potential energy of the nonlinear and linear approximation at maximum rotation, θ_o, are

Nonlinear: $PE = Wh[f_1(\theta_o) - 1]$ [Eq. B-14(a)]

Linear: $PE = 1/2K_R\theta_o^2$ [Eq. B-14(b)]

where

W = Weight of the rigid body
K_R = Linear approximated rotational stiffness
$f_1(\theta_o)$ = Value of $f_1(\theta)$ at angle θ_o

Equating potential energies,

$$K_R = \frac{2Wh}{\theta_o^2}[f_1(\theta_o) - 1] \quad \text{(Eq. B-15)}$$

The effective circular frequency, ω_e, of free vibration is thus

$$\omega_e = \left[\frac{K_R}{I_B}\right]^{1/2} = \left[\frac{2(f_1(\theta_o) - 1)g}{C_I\theta_o^2 h}\right]^{1/2} \quad \text{(Eq. B-16)}$$

On the right-hand side of Eq. (B-12), $f_1(\theta)$ ranges from 1.0 at $\theta = 0$ to $f_1(\theta_o)$ at $\theta = \theta_o$. Similarly, $f_2(\theta)$ ranges from a at $\theta = 0$ to $f_2(\theta_o)$ at $\theta = \theta_o$. Considering that much more time is spent at θ near zero than is spent at $\theta = \theta_o$, the time average values of $f_1(\theta)$ and $f_2(\theta)$ can be approximated by

$$f_{1_A} \approx 1.0 \quad \text{[Eq. B-17(a)]}$$

$$f_{2_A} \approx a \quad \text{[Eq. B-17(b)]}$$

Thus, substituting the linear approximations of Eqs. [B-17(a)] and [B-17(b)] into Eq. (B-12), the approximate linear equation of motion becomes

$$\ddot{\theta} + \omega_e^2 \theta = \frac{-F_H \ddot{x} - a\ddot{y}}{C_I h} \quad \text{(Eq. B-18)}$$

from which

$$C_I h \omega_e^2 \theta_o = [F_H^2(SAH^2) + a^2(SAV^2)]^{1/2} \quad \text{(Eq. B-19)}$$

as long as the horizontal spectral acceleration, SAH, and the vertical spectral acceleration, SAV, are randomly phased with respect to each other. Combining Eqs. (B-16) and (B-19),

$$\frac{2g(f_1(\theta_o) - 1)}{\theta_o} = (F_H)SAH\left[1 + \left(\frac{a(SAV)}{F_H(SAH)}\right)^2\right]^{1/2} \quad \text{(Eq. B-20)}$$

from which Eq. (A-5) is obtained after defining F_V by Eq. [A-5(c)].

The rigid body rocking coefficient of restitution is defined in Ref. [B-9] to be

$$C_R = [1 - C_{MRI}(1 - \cos(2\alpha))] \quad \text{(Eq. B-21)}$$

$$C_{MRI} = \frac{MR^2}{I_B} = \frac{(1 + a^2)}{C_I} \quad \text{(Eq. B-22)}$$

Noting that

$$(1 - \cos(2\alpha)) = 2\sin^2\alpha = \frac{2a^2}{(1 + a^2)} \quad \text{(Eq. B-23)}$$

Eq. (B-21) can be simplified to

$$C_R = \left[1 - \frac{2a^2}{C_I}\right] \quad \text{(Eq. B-24)}$$

The reduction, r, in rotation during a full cycle of response (two rocking impacts) is given by

$$r = C_R^2 \quad \text{(Eq. B-25)}$$

whereas the reduction during a full cycle of response with viscous damping is given by

$$r = e^{-\gamma} \quad \text{(Eq. B-26)}$$

where

$$\gamma = \frac{2\pi\beta}{[1 - \beta^2]^{1/2}} \quad \text{(Eq. B-27)}$$

Equating the reduction, r, from Eqs. (B-25) and (B-26),

$$\gamma = -\ell n(r) = -2\ell n(C_R) \quad \text{(Eq. B-28)}$$

and from Eq. (B-27) the equivalent viscous damping, β_e, is

$$\beta_e = \frac{\gamma}{[4\pi^2 + \gamma^2]^{1/2}} \quad \text{(Eq. B-29)}$$

Thus,

C_R	β_e (%)
0.5	21.6
0.6	16.0
0.7	11.3
0.8	7.1
0.9	3.4
0.95	1.63
0.98	0.64

Refs. [B-2] through [B-6] provide additional information on rigid body sliding and rocking.

SECTION B.3 EXAMPLE PROBLEMS: RIGID BODY ROCKING AND SLIDING

B.3.1 Rigid Body Rocking Example

A rigid body with a height of $H = 84$ in. and width of $B = 36$ in. is subjected to Regulatory Guide 1.60 (Ref. [B-10]) response spectrum shape shown in Figure B-1. Determine the maximum rocking angle, θ_o, and maximum uplift height, δ, as a function of the PGA for this input motion using the approximate rocking method of Section A.2. Assume the vertical spectral acceleration, SAV, is two-thirds of the horizontal spectral acceleration, SAH, at all frequencies.

Assume the mass is uniformly distributed throughout the rigid body. Thus, the center of gravity height, h, and horizontal distance, b, from the rocking corner are

$$h = \frac{H}{2} = 42 \text{ in.} \quad \text{[Eq. B-30(a)]}$$

$$b = \frac{B}{2} = 18 \text{ in.} \quad \text{[Eq. B-30(b)]}$$

From Eqs. [7-2(a)] and [7-2(b)], the aspect ratio, a, and instability angle, α, are

$$a = 0.4286 \quad \text{[Eq. B-31(a)]}$$

$$\alpha = 0.4049 \quad \text{[Eq. B-31(b)]}$$

Assuming the lateral inertial mass, M_L, and the vertical resisting mass, M, both act through the center of gravity and are identical, then from Eqs. [A-5(b)], [A-5(c)], and [A-6(f)],

$$F_H = 1.0 \quad \text{[Eq. B-32(a)]}$$

$$F_V = 1.04 \quad \text{[Eq. B-32(b)]}$$

$$C_I = 1.5782 \quad \text{[Eq. B-32(c)]}$$

Note from F_V that the vertical ground motion has only a very small effect on rocking and is often ignored.

Step 1 Determine the horizontal spectral acceleration capacity, SAH_{CAP}, and uplift displacement, δ, corresponding to any rotation angle, θ_o.

The horizontal spectral acceleration capacity, SAH_{CAP}, is determined from Eq. (A-5), and the uplift displacement, δ, is from

$$\delta = B(\sin \theta_o) \quad \text{(Eq. B-33)}$$

Values of SAH_{CAP} and δ are computed and tabulated in Table B-3 for various rocking angles, θ_o.

Step 2 Determine the effective rocking frequency, f_e, corresponding to any θ_o and effective damping, β_e.

The effective rocking frequency, f_e, is determined from Eq. [A-6(a)] and is tabulated in Table B-3 for various rocking angles, θ_o. The effective damping, β_e, is computed from Eqs. [A-6(b)], [A-6(d)], and [A-6(e)] to be

$$C_R = 0.7672$$

$$\gamma = 0.5299 \quad \text{(Eq. B-34)}$$

$$\beta_e = 0.084 = 8.4\%$$

Step 3 Determine PGA at which SAH_{DEM} determined at frequency, f_e, and damping β_e, equals SAH_{CAP} for various angles, θ_o.

For the Regulatory Guide 1.60 response spectrum shown in Figure B-2, the spectral acceleration demand, SAH_{DEM}, is maximum at the frequency, f_{em}, of

$$f_{em} = 2.5 \text{ Hz} \quad \text{(Eq. B-35)}$$

The rocking angle, θ_{om}, corresponding to f_{em} is obtained from Eq. [A-6(a)], which is most easily solved

TABLE B-3. Rigid Body Rocking

θ_o	$(f_1(\theta_o) - 1)$	SAH_{CAP} (g)	δ (in.)	f_e (Hz)	$\left(\frac{SAH_{DEM}}{PGA}\right)$	PGA_{CAP} (g)
0.0198	0.0083	0.805	0.71	2.50	2.49	0.32
0.02	0.0084	0.805	0.72	2.48	2.47	0.33
0.038	0.0156	0.787	1.37	1.78	1.91	0.41
0.05	0.0202	0.775	1.80	1.54	1.71	0.45
0.10	0.0378	0.727	3.6	1.06	1.28	0.57
0.15	0.0528	0.677	5.4	0.832	1.06	0.64
0.20	0.0652	0.627	7.2	0.694	0.916	0.68
0.25	0.0749	0.576	8.9	0.595	0.813	0.71
0.30	0.0820	0.526	10.6	0.518	0.729	0.72
0.35	0.0863	0.474	12.3	0.456	0.660	0.72
0.40	0.0880	0.423	14.0	0.403	0.600	0.71
0.4049	0.0880	0.413	14.2	0.398	0.594	0.70

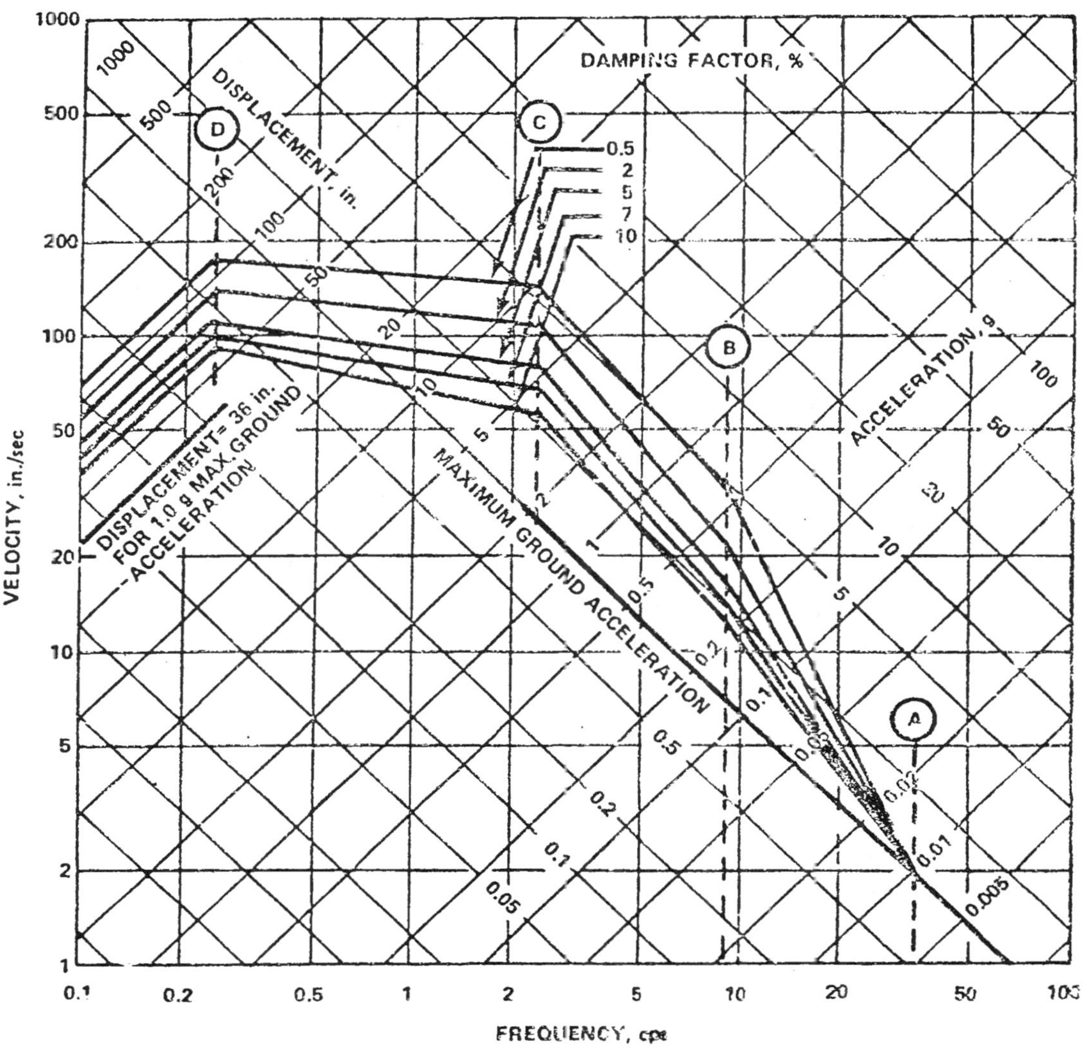

B-2. Horizontal Design Response Spectra, Scaled to 1-g Horizontal Ground Acceleration

by substituting the small angle approximation of Eq. [A-5(c)] for $f_1(\theta_o) - 1$. Thus,

$$\theta_{om} = \frac{2a}{\left[\frac{C_I h}{g}(2\pi f_e)^2\right] + 1} \quad \text{(Eq. B-36)}$$

from which

$$\theta_{om} = 0.0198 \quad \text{(Eq. B-37)}$$

The approximate rocking method of Section A.2 can only be used to predict rocking angles between $\theta_{om} = 0.0198$ and $\alpha = 0.4049$. Above α, rocking becomes unstable. Below θ_{om}, the actual rocking angle, θ_o, will be highly variable, ranging from zero up to θ_{om}, depending on the detailed time-history input.

Therefore, Table B-3 is limited to the range of θ_o between 0.0198 and 0.4049.

For the Regulatory Guide 1.60 response spectrum, at $\beta_e = 8.4\%$,

$$(SAH_{DEM}/PGA) = 2.49\left(\frac{f_e}{2.5\text{Hz}}\right)^{0.780} \quad \text{(Eq. B-38)}$$

$$0.25\text{ Hz} \le f_e \le 2.5\text{ Hz}$$

Values of (SAH_{DEM}/PGA) for the various f_e are also shown in Table B-3. Finally, the PGA_{CAP} are computed from

$$PGA_{CAP} = SAH_{CAP}(SAH_{DEM}/PGA)^{-1} \quad \text{(Eq. B-39)}$$

and shown in Table B-3.

At a PGA of 0.32 g, the rocking angle is only 0.0198 and the uplift displacement is 0.71 in. As the PGA increases above 0.32 g, the rocking angle increases. At a PGA of 0.72 g, the rocking becomes unstable and overturning occurs because θ_c increases unbounded at this angle.

Rocking is initiated when the horizontal spectral acceleration demand, $SAH_{DEM,E}$, at the elastic frequency and elastic damping of the so-called rigid body exceeds

$$SAH_{DEM,E} \geq \frac{ga}{F_H F_V} = 0.41\text{g} \qquad \text{(Eq. B-40)}$$

If the so-called rigid body is truly a rigid body, then

$$SAH_{DEM,E} = PGA \qquad \text{[Eq. B-41(a)]}$$

and rocking will not initiate until $PGA = 0.41$ g. Once initiated, the rocking angle will immediately increase to $\theta_o = 0.038$ for which $PGA_{CAP} = 0.41$ g in Table B-3.

However, if the elastic frequency is 15 Hz and damping is 5%, then

$$SAH_{DEM,E} = 1.79\ PGA \qquad \text{[Eq. B-41(b)]}$$

so that rocking will initiate at $PGA = 0.23$ g. Between $PGA = 0.23$ g and 0.32 g, the rocking angle will be highly uncertain between zero and $\theta_{om} = 0.0198$, which is still a very small rocking angle.

B.3.2 Rigid Body Sliding Example

Sliding of an unanchored rigid body is resisted by a coefficient of friction, μ, of

$$\mu = 0.40 \qquad \text{(Eq. B-42)}$$

and is subjected to the Regulatory Guide 1.60 (Ref. [B-10]) response spectrum shape shown in Figure B-2 in each of two orthogonal horizontal directions. Determine the sliding displacement, δ_S, as a function of the PGA for this input motion using the approximate sliding method of Section A.1. Assume the vertical acceleration, A_V, is two-thirds of the horizontal PGA.

From Eqs. (A-1) and (A-2),

$$\mu_e = 0.4\left[1 - 0.267\,\frac{PGA}{g}\right] \qquad \text{(Eq. B-43)}$$

$$c_S = 0.8g\left[1 - 0.267\,\frac{PGA}{g}\right] \qquad \text{(Eq. B-44)}$$

For the Regulatory Guide 1.60 response spectrum shown in Figure B-2, the maximum spectral acceleration demand occurs at the frequency, f_{em}, of

$$f_{em} = 2.5\text{ Hz} \qquad \text{(Eq. B-45)}$$

For frequencies below 2.5 Hz, the 10% damped horizontal spectral acceleration demands, SA_{H_1} and SA_{H_2}, are given by

$$SA_{H_1} = SA_{H_2} = 2.28PGA\left(\frac{f_{eS}}{2.5\text{Hz}}\right)^{0.766} \qquad \text{[Eq. B-46(a)]}$$

$$0.25\text{ Hz} \leq f_e\ 2.5\text{ Hz}$$

$$SA_{H_1} = SA_{H_2} = 0.391PGA\left(\frac{f_{eS}}{0.25\text{Hz}}\right)^{2} \qquad \text{[Eq. B-46(b)]}$$

$$f_e \leq 0.25\text{ Hz}$$

Thus, from Eq. (A-4), the 10% damped vector spectral acceleration, SA_{VH}, is

$$SA_{VH} = 2.46PGA\left(\frac{f_{eS}}{2.5\text{Hz}}\right)^{0.766} \qquad \text{[Eq. B-47(a)]}$$

$$0.25\text{ Hz} \leq f_e \leq 2.5\text{ Hz}$$

$$SA_{VH} = 0.421PGA\left(\frac{f_{eS}}{0.25\text{Hz}}\right)^{2} \qquad \text{[Eq. B-47(b)]}$$

$$f_e \leq 0.25\text{ Hz}$$

For any given PGA value, determine f_{eS} for which SA_{VH} equals c_S from Eq. (B-44). Lastly, compute the sliding displacement, δ_S, from Eq. (A-3). Results for this example problem are tabulated in Table B-4.

Sliding is initiated when the vector horizontal spectral acceleration demand, $SA_{VH,E}$, at the elastic

TABLE B-4. Rigid Body Sliding

PGA (g)	C_S (g)	f_{eS} (Hz)	δ_S (in.)
0.30	0.736	2.50	1.15
0.34	0.728	2.08	1.64
0.40	0.715	1.65	2.57
0.50	0.693	1.18	4.85
0.60	0.672	0.895	8.20
0.70	0.651	0.702	12.9
0.80	0.629	0.564	19.3
0.90	0.608	0.463	27.8
1.00	0.587	0.385	38.7

frequency and elastic damping of the so-called rigid body exceeds

$$SA_{VH,E} \geq \mu_e g = 0.4\ g - 0.107\ PGA \quad \text{(Eq. B-48)}$$

If the so-called rigid body is truly a rigid body, then

$$SA_{VH,E} = 1.08\ PGA \quad \text{[Eq. B-49(a)]}$$

and sliding will not initiate until $PGA = 0.34$ g. Once initiated, the sliding displacement will jump to 1.64 in., as shown in Table B-4 for a *PGA* of 0.34 g.

However, if the elastic frequency is 15 Hz and damping is 5%, then

$$SA_{VH,E} = 1.93\ PGA \quad \text{[Eq. B-49(b)]}$$

so that sliding will initiate at $PGA = 0.20$ g. Between $PGA = 0.20$ g and 0.30 g, the sliding displacement will be highly uncertain between 0 in. and 1.15 in.

REFERENCES FOR APPENDIX B.0

[B-1] Reed, J.W., Kennedy, R.P., and Lashkari, B. (1993). "Analysis of high frequency seismic effects." *EPRI TR-102470*, Electric Power Research Inst., Palo Alto, Calif.

[B-2] Aslam, M., Godden, W.G., and Scallse, D.T. (1980). "Earthquake rocking response of rigid bodies." *J. Struct. Engrg.* vol. 106 (Feb.).

[B-3] Blume, J.A. (1960). "A reserve energy technique for the earthquake design and rating of structures in the inelastic range." In *Proc., 2nd World Conf. on Earthquake Engineering,* pp. 1061–1083. Science Council of Japan, Tokyo.

[B-4] Ishiyama, Y. (1983). "Motions of rigid bodies and criteria for overturning by earthquake excitations." In *Proc., 3rd South Pacific Regional Conf. on Earthquake Engineering,* Wellington, New Zealand. [Also in *Earthquake Engrg. and Struct. Dynamics*, 10, 635–650 (1982).]

[B-5] Housner, G.W. (1963). "The behavior of inverted pendulum structures during earthquakes." *Bull. Seismological Society of America*, 53(2), 403–417.

[B-6] Milne, J. (1881). "Experiments in observational seismology." *Trans., Seismological Society of Japan*, vol. 3 (Jan.–Dec.).

[B-7] Newmark, N.M. (1965). "Effects of earthquakes on dams and embankments." Fifth Rankine Lecture, Geotechnique.

[B-8] Newmark, N.M., and Rosenblueth, E. (1971). "Fundamentals of earthquake engineering." Prentice-Hall, Englewood Cliffs, N.J.

[B-9] Yim, S.C.-S., Chopra, A.K., and Penzien, J. (1980). "Rocking response of rigid blocks to earthquakes." *Earthquake Engrg. and Struct. Dynamics*, 8, 565–580.

[B-10] U.S. Nuclear Regulatory Commission. (1973). "Design response spectra for seismic design of nuclear power plants." *Regulatory Guide 1.60*, Washington, D.C.

COMMENTARY TO AMERICAN SOCIETY OF CIVIL ENGINEERS STANDARD 43-05

(This Commentary is not a part of the ASCE Standard *Seismic Design Criteria for Structures, Systems, and Components in Nuclear Facilities.* It is included for information purposes.)

This Commentary consists of explanatory and supplementary material designed to assist local building code committees and regulatory authorities in applying the recommended requirements. In some cases, it will be necessary to adjust specific values in the Standard to location conditions; in others, a considerable amount of detailed information is needed to put the provisions into effect. This Commentary provides a place for supplying material that can be used in these situations and is intended to create a better understanding of the recommended requirements through brief explanations of the reasoning employed in arriving at them.

The sections of the Commentary are numbered to correspond to the sections of the Standard to which they refer. Since it is not necessary to have supplementary material for every section in the Standard, there are gaps in the numbering in the Commentary.

SECTION C1.0 INTRODUCTION

SECTION C1.1 OVERVIEW OF THE SEISMIC DESIGN CRITERIA

The objective of the seismic design criteria is to safeguard against the accidental release of nuclear material by requiring SSCs in seismic regions to be seismically robust. Thus, these criteria focus on cast-in-place reinforced concrete and structural steel systems. Stringent detailing requirements are specified for these structural systems to ensure seismic ruggedness with a high degree of reliability.

Structural systems with a demonstrated poor seismic performance record—such as (1) unreinforced masonry, or (2) precast concrete using conventional gravity connections—are purposely omitted and shall not be utilized in the primary seismic load path.

These criteria are not intended to discourage the use of any structural system with reliable seismic performance. Guidance is provided to develop criteria for seismically rugged structural systems that ensure the same degree of safety as the structural systems explicitly considered.

SECTION C1.2 USE OF ASCE STANDARD 43-05 WITH OTHER CODES AND STANDARDS

Representative applications of the graded approach provided by specifying the SDC and the Limit State are shown in Table C1-1.

The earthquake provisions in the 2003 International Building Code and ASCE 7-02 are based primarily on the 1997 edition of the NEHRP Recommended Provisions for Seismic Regulations for New Buildings and Other Structures (referred to here as "the Provisions"). The purposes of the Provisions are stated as follows:

- To provide minimum design criteria for structures appropriate to their primary function and use considering the need to protect the health, safety, and welfare of the general public by minimizing the earthquake-related risk to life, and
- To improve the capability of essential facilities and structures containing substantial quantities of hazardous materials to function during and after design earthquakes.

The design earthquake ground motion levels specified in the Provisions and the IBC could result in both structural and nonstructural damage. For most structures designed and constructed according to the Provisions and IBC, structural damage from the design earthquake ground motion would be repairable although perhaps not economically so. For essential facilities, it is expected that the damage from the design earthquake ground motion would not be so severe as to preclude continued occupancy and function of the facility. The actual ability to accomplish these goals depends on a number of factors, including the structural framing type, configuration, materials, and as-built details of construction. For ground motions larger than the design levels, the intent of the Provisions and the IBC is that there should be a low likelihood of structural collapse.

The Provisions were developed with the intent to provide uniform levels of performance for structures, depending on their occupancy and use and the risk to society inherent in their failure. The Provisions establish a series of Seismic Use Groups that are used to categorize structures based on the specific SDC. It is the intent of the Provisions that a uniform margin of failure to meet the seismic design criteria be provided for all structures within a given Seismic Use Group.

TABLE C1-1. Representative Applications of the Graded Approach

	Limit State			
SDC	Large Permanent Distortion A	Moderate Permanent Distortion B	Limited Permanent Distortion C	Essentially Elastic Behavior D
1	ASCE 7 (SG I) DOE PC 1 Systems interactions	ASCE 7 (SG II)	ASCE 7 (SG III)	Not addressed
2	ASCE 7 (SG II) Systems interactions	ASCE 7 (SG III) DOE PC 2	Not addressed	Not addressed
3	Systems interactions		DOE PC 3 Many DOE SSCs	
4	Systems interactions			
5	Systems interactions		DOE PC 4 Near NRC NPP	Similar to modern NRC NPP

Note:
NPP = Nuclear power plant; SG = Seismic Use Group.

In past editions of the Provisions, it was stated that the seismic hazards around the nation were defined at a uniform 10% probability of exceedance in 50 yr (although rigorous probabilistic seismic hazard studies were not performed to support the statements), and the design requirements were based on assigning a structure to a Seismic Hazard Exposure Group and a Seismic Performance Category. While this approach provided for a uniform likelihood throughout the nation that the design ground motion would not be exceeded, it did not provide for a uniform margin of failure for structures designed for that ground motion. The reason for this is that the rate of change of earthquake ground motion versus likelihood is not constant in different regions of the United States.

Review of rigorous modern probabilistic seismic hazard results, including the maps prepared by the U.S. Geological Survey (USGS) to support the development of the 1997 Provisions, indicates that the rate of change of ground motion versus probability is not constant throughout the United States. For example, the ground motion difference between the 10% probability of exceedance and 2% probability of exceedance in 50 yr is typically smaller than the difference between the two probabilities in less active areas, such as the eastern or central United States. Figure C1-1 plots the 0.2-s spectral acceleration normalized at 2% probability of exceedance in 50 yr versus the annual frequency of exceedance. Figure C1-1 shows that in coastal California, the ratio between the 0.2-s spectral acceleration for the 2% and 10% probabilities of exceedance in 50 yr is about 1.5, whereas in other parts of the United States, the ratio varies from 2.0 to 5.0.

As stated earlier, the goals of the Provisions are basically to protect the life safety of occupants and to provide for a low likelihood of structural collapse for ground motions larger than the design levels. In order to accomplish this—considering the differences in seismic hazard in the nation—ground motion hazards are defined in terms of maximum considered earthquake ground motions (i.e., earthquake ground motion levels for which a low likelihood of structural collapse is desired). The design earthquake ground motions are based on a lower bound estimate of the margin against collapse inherent in structures designed to the Provisions. This lower bound was judged, based on experience, to be about a factor of 1.5 in ground motion. Consequently, the design earthquake ground motion is selected at a ground shaking level that is 1/1.5 (or 2/3) of the maximum considered earthquake ground motion. Another way of stating this is, if a structure is designed for a level of ground shaking in accordance with the Provisions, then that structure shall be able to resist a level of shaking 1.5 times the design level with a low likelihood of collapse.

For most regions of the United States, the maximum considered earthquake ground motion is defined with a uniform likelihood of exceedance of 2% in 50 yr (return period of about 2500 yr). While stronger ground motion than this can occur, it was judged that it would be economically impractical to design for such rare ground motions and the selection of the 2% in 50 yr likelihood as the maximum considered earthquake ground motion would result in acceptable levels of seismic safety for the nation.

In regions of high seismicity, such as coastal California, the seismic hazard is typically controlled by large-magnitude events occurring on a limited number of well-defined fault systems. Ground shaking calculated at a 2% in 50 yr likelihood would be much larger than that which would be expected based on the characteristic magnitudes of earthquakes on these known active faults. This is because these major active faults can produce characteristic earthquakes every few hundred yr. For these regions it is considered more appropriate to directly determine maximum considered earthquake ground motion based on the characteristic earthquakes of these defined faults. In order to provide for an appropriate level of conservatism in the design process, when this approach to calculation of the maximum considered earthquake ground motion is used, the median estimate of ground motion resulting from the characteristic event is multiplied by 1.5.

The Provisions and the IBC assign structures to one of three Seismic Use Groups, as follows:

- Seismic Use Group I structures (such as a small office building) are those not assigned to Seismic Use Groups II or III.
- Seismic Use Group II structures (such as a public assembly structure with a capacity greater than 300 persons, and schools) are those that have a substantial public hazard due to occupancy or use.
- Seismic Use Group III structures (such as fire, rescue, and police stations; hospitals; emergency preparedness centers) are those having essential facilities that are required for post-earthquake recovery and those containing substantial quantities of hazardous substances.

The reason for assigning structures to the Seismic Use Groups is to determine the Occupancy Importance Factors, which is used as part of determining the earthquake loads for the design of the structures. The

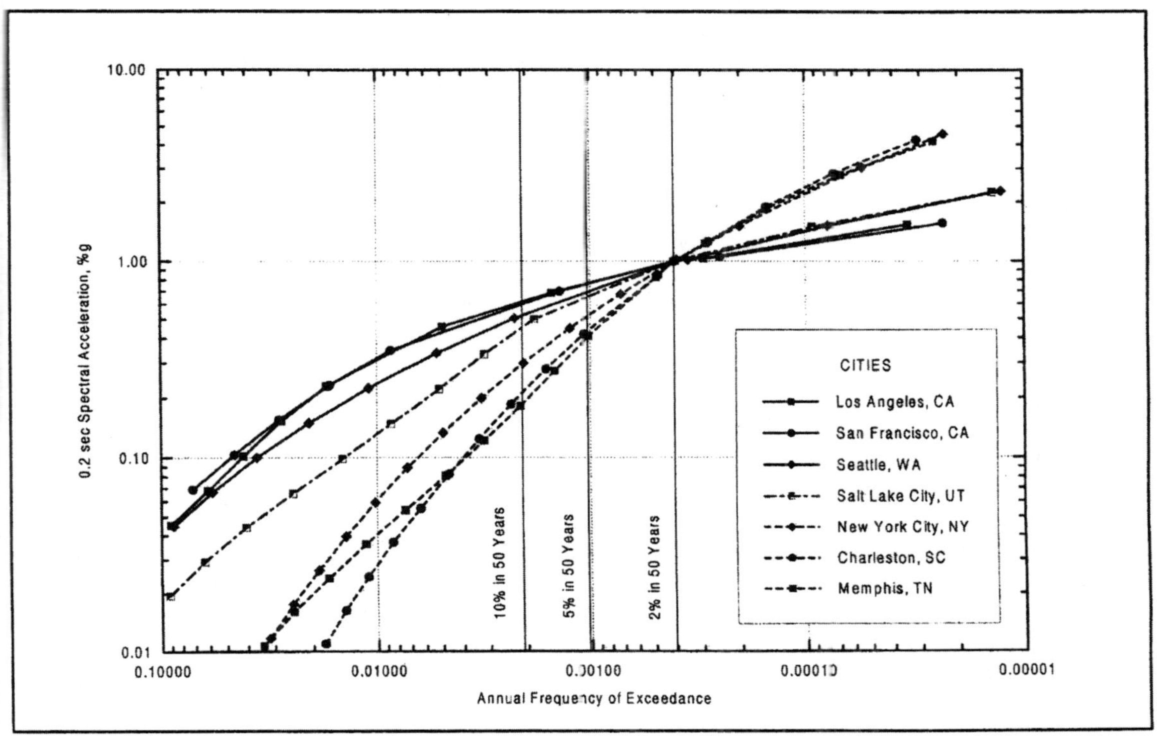

C1-1. Relative Seismic Hazard at Selected West Coast and East/Central U.S. Sites for 0.2-s Spectral Response Acceleration

Occupancy Importance Factors for the Seismic Use Groups are as follows:

- Seismic Use Group I: 1.0
- Seismic Use Group II: 1.25
- Seismic Use Group III: 1.5

For example, a hospital would be designed for earthquake loads equal to 1.5 times the loads for a small office building.

In addition to the Seismic Use Groups, the Provisions and the IBC define six SDCs (A, B, C, D, E, and F) based on the level of the design earthquake ground motions, with SDC-A being for the lowest level of ground motion and SDC D, E, and F being the highest levels of design ground motions. SDCs are used in the Provisions and the IBC to determine permissible structural systems, limitations on height and irregularity, those components of the structure that must be designed for seismic resistance, detailing requirements, and the types of lateral force analysis that must be performed. The requirements become more rigorous and stringent as the SDC goes from A to F.

The U.S. DOE has defined Performance Categories (PC) to define the seismic hazard and the seismic analyses/design criteria to use for the SSCs in the DOE facilities, which range from conventional facilities to research reactors. Table C1-1 shows where the DOE PCs fit into the graded approach being defined by this Standard. SSCs in the DOE facilities are currently placed into DOE PCs ranging from 0 to 4, as follows:

DOE PC 0 No criteria
DOE PC 1 Building code criteria for conventional facilities
DOE PC 2 Building code criteria for essential facilities
DOE PC 3 Intermediate criteria above building codes and below commercial nuclear power plant criteria
DOE PC 4 Approaching criteria used for commercial nuclear power plants

SECTION C1.3 ALTERNATIVE METHODS TO MEET INTENT OF THIS STANDARD

The DBE ground motion is defined in terms of a DRS defined by Eq. (2-1). The Design Factor, *DF*, used in Eq. (2-1) to define the DRS is aimed at achieving the Target Performance Goal annual frequencies defined in Table 2-1, as long as the seismic demand and structural capacity evaluations have sufficient conservatism to achieve both of the following:

1. Less than about a 1% probability of unacceptable performance for the DBE ground motion, and
2. Less than about a 10% probability of unacceptable performance for a ground motion equal to 150% of the DBE ground motion.

Therefore, alternative methods that are aimed at achieving the above-specified level of conservatism are acceptable.

The Standard is based on achieving both probability goals, which represent two points on the underlying fragility curve. Having these two probability goals allows the target probabilities to be achieved with less possibility of unconservatism. The work required to demonstrate that both goals are achieved when alternative methods are used is only slightly greater than showing that one of the two goals is achieved.

Seismic fragility (conditional probabilities of failure versus ground motion levels) is typically defined as being lognormally distributed so that it can be fully described in terms of a seismic margin factor, F_{PF} (factor applied to the DBE ground motion) corresponding to a conditional failure probability, P_{FC}, and an estimate of the failure variability (logarithmic standard deviation β). The two target levels of conservatism defined above result in the following seismic margin factors—$F_{1\%}$, $F_{5\%}$, $F_{10\%}$ and $F_{50\%}$—corresponding to a 1%, 5%, 10%, and 50% conditional probability of failure (unacceptable behavior), respectively:

β	$F_{1\%}$	$F_{5\%}$	$F_{10\%}$	$F_{50\%}$
0.30	1.10	1.35	1.50	2.20
0.40	1.00	1.31	1.52	2.54
0.50	1.00	1.41	1.69	3.20
0.60	1.00	1.50	1.87	4.04

For a logarithmic standard deviation less than 0.39, the second of the two conditional failure probability goals controls the fragility. For β greater than 0.39, the first goal controls. By specifying both goals, the following margins are achieved:

$F_{1\%}$ ≥ 1.0

$F_{5\%}$ ≥ 1.3

$F_{10\%}$ ≥ 1.5

$F_{50\%}$ Increases with increasing β

These minimum margins are sufficient to reasonably achieve the Target Performance Goals, as shown in the sections that follow.

C1.3.1 Expected Factors of Safety Achieved by Seismic Acceptance Criteria

C1.3.1.1 Introduction

In this Standard, strengths are specified in terms of the ACI code ultimate strengths, the AISC code LRFD Limit State strengths including the code specified strength reduction factors (ϕ), and the ASME code Service Level D strengths. The seismic demand is specified in terms of ASCE 4 requirements. For ductile failure modes, appropriately conservative inelastic energy absorption factors, F_μ, are specified within this Standard in Table 5-1.

In this section, the resulting strength, seismic demand and nonlinear factors of conservatism are first estimated and then combined to obtain an overall estimate of the factor of safety achieved by the seismic acceptance criteria specified in this Standard.

C1.3.1.2 Estimation of Median Conservatism Introduced by Standard Seismic Acceptance Criteria

The median seismic capacity, $C_{50\%}$, can be estimated from

$$C_{50\%} = \frac{S_{50\%}}{D_{50\%}} F_{\mu_{50\%}} DBE \qquad \text{(Eq. C1-1)}$$

where

$S_{50\%}$ = Median estimates of the component seismic strength
$D_{50\%}$ = Seismic demand for a specified DBE input
$F_{\mu_{50\%}}$ = Inelastic energy absorption (nonlinear) factor

In turn, the standard seismic capacity, C_{STD}, is given by

$$C_{STD} = \frac{S_{STD}}{D_{STD}} F_{\mu_{STD}} DBE \qquad \text{(Eq. C1-2)}$$

where

S_{STD} = Deterministic strength
D_{STD} = Demand
$F_{\mu_{STD}}$ = Nonlinear factors

defined in accordance with this Standard. Defining R_S, R_D, and R_N as the median conservatism ratios associated with this Standard, then

$$S_{50\%} = R_S\, S_{STD}$$
$$D_{50\%} = D_{STD}/R_D \qquad \text{(Eq. C1-3)}$$
$$F_{\mu_{50\%}} = R_N\, F_{\mu_{STD}}$$

and

$$C_{50\%} = R_C\, C_{STD} \qquad \text{(Eq. C1-4)}$$
$$R_C = R_S\, R_D\, R_N \qquad \text{(Eq. C1-5)}$$

where R_C is the overall median conservatism ratio associated with this Standard's acceptance criteria. The ratios R_S, R_D, and R_N are estimated in the following three subsections.

C1.3.1.2.1 Median Strength Conservatism Ratio

Based on a review of median capacities from past seismic probabilistic risk assessment studies versus U.S. code-specified ultimate strengths for a number of failure modes, it is judged that for ductile failure modes when the conservatism of material strengths, code strength equations, and seismic strain-rate effects are considered, the code ultimate strengths have at least a 98% probability of exceedance. For low-ductility failure modes, an additional factor of conservatism of about 1.33 is typically introduced.

Thus,

$$\begin{aligned} &\text{(Ductile)} && R_S = e^{2.054\beta_S} \\ &\text{(Low ductility)} && R_S = 1.33e^{2.054\beta_S} \end{aligned} \qquad \text{(Eq. C1-6)}$$

where β_S is the strength logarithmic standard deviation (typically in the range of 0.2 to 0.4), and 2.054 is the standardized normal variable for 2% NEP.

C1.3.1.2.2 Median Demand Conservatism Ratio

Seismic demands are computed in accordance with the requirements of ASCE 4 except that median input spectral amplifications are used instead of median-plus-one standard deviation amplification factors. When both are anchored to the same average spectral acceleration computed over a broad frequency range of interest, such as 3 Hz to 8 Hz, the ratio of median-plus-one standard deviation to median spectral acceleration amplification factor averages about 1.22. In addition, as noted in its preface, ASCE 4 is aimed at achieving about a 10% probability of the actual seismic response exceeding the computed response, given the occurrence of the DBE. Thus, the median demand ratio, R_D, can be estimated from

$$R_D = \frac{e^{1.282\beta_D}}{1.22} \qquad \text{(Eq. C1-7)}$$

where β_D is the seismic demand logarithmic standard deviation for a specified seismic input (typically in the range of 0.2 to 0.4).

TABLE C1-2. Nominal Factor of Safety, $F_{N1\%}$

Strength Variability (β_S)	Demand Variability (β_D)	Low-ductility Failure Modes	Ductile Failure Modes	
			$\beta_N = 0.2$	$\beta_N = 0.4$
0.2	0.2	1.10	0.99	0.99
	0.3	1.04	0.97	1.00
	0.4	0.97	0.92	0.99
0.3	0.2	1.12	1.04	1.08
	0.3	1.11	1.04	1.11
	0.4	1.05	1.00	1.10
0.4	0.2	1.13	1.07	1.15
	0.3	1.13	1.08	1.19
	0.4	1.11	1.07	1.20

C1.3.1.2.3 Median Nonlinear Conservatism Ratio

In this Standard, the nonlinear factor is aimed at about the 5% NEP level. Thus, for ductile failure modes, the median nonlinear factor ratio, R_N, should be

$$\text{Ductile} \quad R_N = e^{1.645\beta_N} \qquad \text{[Eq. C1-8(a)]}$$

where

β_N = Logarithmic standard deviation for the nonlinear factor (typically in the range of 0.2 to 0.4 for ductile failure modes)

1.645 = Standardized normal variable for 5% NEP

However, for low-ductility (brittle) failure modes, no credit is taken for a nonlinear factor; that is,

$$\text{Brittle} \quad \begin{aligned} F_{\mu 50\%} &= 1.0 \\ R_N &\approx 1.0 \end{aligned} \qquad \text{[Eq. C1-8(b)]}$$

C1.3.1.3 Resulting Capacity Conservatism

Combining Eqs. (C1-5) through (C1-8), the median capacity ratio, R_C, is estimated to be

(Ductile failures)

$$R_C = 0.82e^{2.054\beta_S + 1.282\beta_D + 1.645\beta_N} \qquad \text{(Eq. C1.3-9)}$$

(Low ductility) $R_C = 1.09e^{2.054\beta_S + 1.282\beta_D}$

and

$$C_{1\%} = R_C\, C_{STD}\, e^{-2.326\beta} \qquad \text{(Eq. C1-10)}$$

$$\beta = [\beta_S^2 + \beta_D^2 + \beta_N^2]^{1/2} \qquad \text{(Eq. C1-11)}$$

The resulting nominal factor of safety, $F_{N1\%}$, against a 1% conditional probability of failure is then given by

$$F_{N1\%} = \frac{C_{1\%}}{C_{STD}} = R_C e^{-2.326\beta} \qquad \text{[Eq. C1-12(a)]}$$

TABLE C1-3. Nominal Factor of Safety, $F_{N10\%}$

Strength Variability (β_S)	Demand Variability (β_D)	Low-ductility Failure Modes	Ductile Failure Modes	
			$\beta_N = 0.2$	$\beta_N = 0.4$
0.2	0.2	1.48	1.42	1.64
	0.3	1.52	1.49	1.76
	0.4	1.54	1.53	1.84
0.3	0.2	1.64	1.60	1.89
	0.3	1.72	1.69	2.03
	0.4	1.77	1.76	2.15
0.4	0.2	1.80	1.78	2.15
	0.3	1.91	1.90	2.32
	0.4	2.00	2.00	2.46

Similarly, the nominal factor of safety, $F_{N10\%}$, against a 10% conditional probability of failure is given by

$$F_{N10\%} = \frac{C_{10\%}}{C_{STD}} = R_C e^{-1.282\beta} \quad \text{[Eq. C1-12(b)]}$$

Table C1-2 presents $F_{N1\%}$ for typical values of β_S, β_D, and β_N. It can be seen that, over this entire range of β values,

$$F_{N1\%} \approx 1.0 \quad \text{(Eq. C1-13)}$$

with $F_{N1\%}$ ranging from 0.93 to 1.20 with a median value of 1.07. Table C1-3 presents $F_{N10\%}$ for typical values of β_S, β_D, and β_N. It also can be seen that

$$F_{N10\%} \approx 1.5 \quad \text{(Eq. C1-14)}$$

Thus, both Eqs. (C1-13) and (C1-14) are satisfied by the seismic acceptance criteria presented in this Standard.

REFERENCES FOR SECTION C1.0

[C1-1] ANS. (2005). "American National Standard for design categorization of nuclear facility structures, systems, and components for natural phenomena hazards." *ANSI/ANS 2.26*, La Grange Park, Ill.

[C1-2] ANS. (2005). "Site characterization requirements for natural phenomena hazards at nuclear materials facilities sites." *ANSI/ANS 2.27*, La Grange Park, Ill.

[C1-3] ANS. (2005). "Probabilistic analysis of natural phenomena hazards for nuclear materials facilities." *ANSI/ANS 2.29*, La Grange Park, Ill.

[C1-4] ASME. (2000). *Quality Assurance Requirements for Nuclear Facility Applications, NQA-1* New York.

[C1-5] International Code Council, Inc. (2003). *International Building Code 2003*, Whittier, Calif.

[C1-6] FEMA. (1997). NEHRP provisions for seismic regulations for new buildings and other structures Washington, D.C.

SECTION C2.0 EARTHQUAKE GROUND MOTION

SECTION C2.2 DEVELOPMENT OF DESIGN BASIS EARTHQUAKE GROUND MOTION

C2.2.1 Horizontal Ground Motion

C2.2.1.1 Introduction

Figure C2-1 presents normalized seismic hazard curves for eleven representative sites expressed in terms of mean annual frequency of exceedance versus normalized spectral acceleration. Spectral accelerations are normalized to their value at an exceedance frequency of 10^{-4} to enable an easy comparison of the hazard curve shapes. The nine sites labeled "EUS" are representative of central and eastern U.S. sites ranging from Arkansas to Illinois to Maine to Georgia. The California site is near Santa Maria, Calif., which has high natural frequency ground motion dominated by active nearby faults and low-frequency ground motion dominated by large events on the more distant San Andreas Fault. The Washington site is in south-central Washington with high natural frequencies also dominated by local earthquakes and low natural frequencies dominated by a large subduction zone earthquake.

Several general observations can be made for the normalized hazard curves shown in Figure C2-1:

1. The hazard curves are much steeper for the California and Washington sites than for the EUS sites.
2. Hazard curve slopes are slightly steeper at 10 Hz than at 1 Hz, particularly for the EUS sites.
3. When plotted on a log-log plot, the hazard curves can be approximated as being linear at least over any 10-fold exceedance frequency range.

Based on these observations, Ref. [C2-1] has suggested an approach to achieving risk-consistent seismic design criteria even with widely varying hazard curve slopes. This approach requires that a UHRS be scaled by a Design Factor, DF, so as to obtain a uniform risk-consistent DRS, SA_{DBE}; that is,

$$DRS = DF \times UHRS \quad \text{(Eq. C2-1)}$$

The Design Factor, DF, must be a function of both of the following:

1. A probability ratio, R_P, defined by

$$R_P = \frac{H_D}{P_F} \quad \text{(Eq. C2-2)}$$

where

H_D = Annual exceedance frequency at which the UHRS is defined

P_F = Target annual frequency of unacceptable seismic performance

2. A hazard curve slope ratio, A_R, defining the change in input ground motion corresponding to a

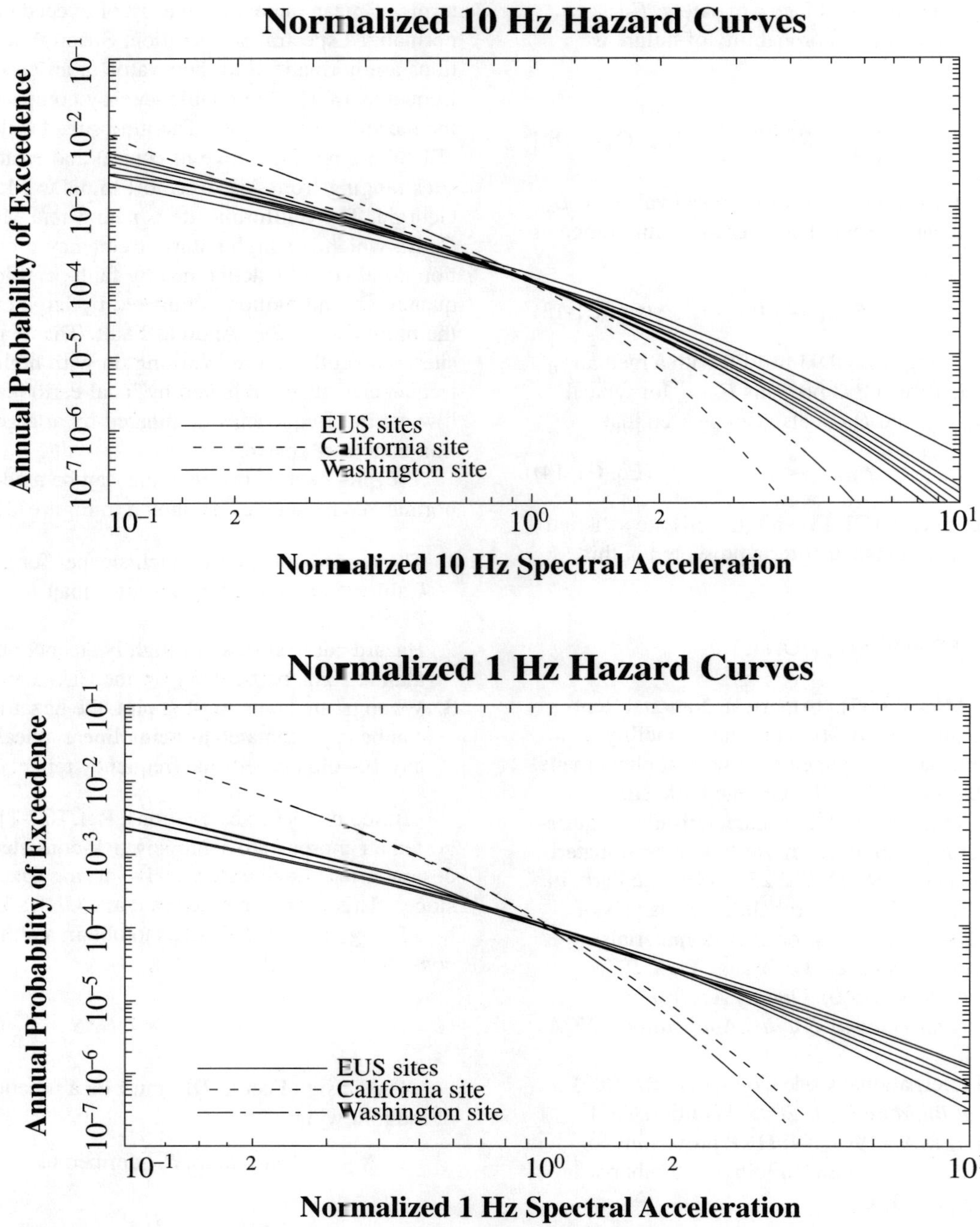

C2-1. Seismic Hazard Curves Normalized by the Spectral Acceleration Value Corresponding to 10^{-4} Annual Probability

10-fold change in exceedance frequency. This slope ratio can be defined by

$$A_R = \frac{SA_{0.1H_D}}{SA_{H_D}} \qquad \text{(Eq. C2-3)}$$

where

SA_{H_D} = Spectral acceleration at the exceedance frequency, H_D
$SA_{0.1H_D}$ = Spectral acceleration at $0.1H_D$

Based on the derivation and results presented in Ref. [C2-1], which are summarized and amplified on in this Standard, the appropriate Design Factors, DF, can be derived for any set of R_P and A_R ratios corresponding to a specified level of conservatism in the seismic design criteria.

C2.2.1.2 Design Factor Derivation

As previously noted, over any 10-fold difference in exceedance frequencies, seismic hazard curves may be approximated by

$$H(a) = K_I a^{-K_H} \qquad \text{(Eq. C2-4)}$$

where

$H(a)$ = Annual frequency of exceedance of ground motion level, a
K_I = Appropriate constant
K_H = Slope parameter defined by

$$K_H = \frac{1}{\log(A_R)} \qquad \text{(Eq. C2-5)}$$

in which A_R is the ratio of ground motions corresponding to a 10-fold reduction in exceedance frequency.

In order to compute the probability ratio, R_P, corresponding to any specified seismic design criteria, one must also define a mean seismic fragility curve for a component resulting from the usage of these seismic criteria. This mean seismic fragility curve describes the conditional probability of an unacceptable performance versus the ground motion level. This mean fragility curve is most often defined as being lognormally distributed and is expressed in terms of two parameters: (1) a median capacity level and (2) a composite logarithmic standard deviation, β, (see Ref. [C2-2] for further amplification). The logarithmic standard deviation, β, will generally lie within the range of 0.3 to 0.5 for structures and equipment mounted at ground level. For equipment mounted high in structures, β will generally lie within the range of 0.4 to 0.6.

For any component, the mean probability, P_F, of unacceptable performances is obtained by convolution of the seismic hazard and fragility curves expressed by

$$P_F = -\int_0^{+\infty} \left(\frac{dH_{(a)}}{da}\right) P_{F/a}\, da \qquad \text{(Eq. C2-6)}$$

where $P_{F/a}$ is the conditional probability of failure given the ground motion level, a, which is defined by the mean fragility curve.

Approximating the seismic hazard curve between UHRS exceedance frequency, H_D, and the failure probability, P_F, by Eq. (C2-4), Ref. [C2-1] shows that

$$R_P = \frac{H_D}{P_f} = (F_P)^{K_H} e^f \qquad \text{(Eq. C2-7)}$$

$$f = X_P K_H \beta - ½(K_H \beta)^2 \qquad \text{(Eq. C2-8)}$$

where

F_P = Factor of safety (seismic margin) between the DBE ground motion input and the component seismic capacity, C_P, associated with the conditional failure probability, $P_{F/a}$
X_P = Standardized normal variant associated with the failure probability, $P_{F/a}$

From Eq. (C2-7), the factor of safety, F_P, required to achieve a required probability ratio, R_P, is

$$F_P = [R_P e^{-f}]^{1/K_H} \qquad \text{(Eq. C2-9)}$$

The seismic acceptance criteria specified in this Standard assume that both of the following nominal factors of safety, F_N, are achieved:

$F_{N1\%} \geq 1.0$ against a 1% conditional probability of failure [Eq. C2-10(a)]

$F_{N10\%} \geq 1.5$ against a 10% conditional probability of failure [Eq. C2-10(b)]

See Section C1 3.1 for justification that both of these nominal factors of safety are achieved. Based on Eq. [C2-10(a)], the UHRS would be multiplied by a Design Factor, DF_a, given by

$$DF_a = F_{P1\%} \qquad \text{[Eq. C2-11(a)]}$$

where $F_{P1\%}$ is computed using the 1% standardized normal variant, $X_{P1\%} = 2.326$, in Eq. (C2-8).

TABLE C2-1. Probability Ratios, R_P, Corresponding to SDC 3, 4, and 5

SDC	H_D	P_F	$R_P = H_D/P_F$
3	4×10^{-4}	1×10^{-4}	4
4	4×10^{-4}	4×10^{-5}	10
5	1×10^{-4}	1×10^{-5}	10

Alternatively, based on Eq. [C2-10(b)], the UHRS would be multiplied by a Design Factor, DF_b, given by

$$DF_b = F_{P10\%} / 1.5 \qquad \text{[Eq. C2-11(b)]}$$

where $F_{P10\%}$ is computed using the 10% standardized normal variant, $X_{P10\%} = 1.282$, in Eq. (C2-8). For logarithmic standard deviations, β, ranging from 0.3 to 0.6, the ratios of (DF_b / DF_a) are as follows:

β	DF_b / DF_a
0.3	0.91
0.4	1.01
0.5	1.12
0.6	1.25

For logarithmic standard deviations ranging from 0.3 to 0.6, the nominal factors of safety, F_N, specified by Eqs. [C2-10(a)] and [C2-10(b)] produce similar required Design Factors, DF_a and DF_b. Since the seismic acceptance criteria specified in this Standard assume that both Eqs. [C2-10(a)] and [C2-10(b)] are achieved, the required Design Factor, DF, should be the lesser of DF_a or DF_b. However, because of some concerns that Eq. [C2-10(a)] may not be met in a few situations where the component seismic capacity has large variability (i.e., high β), and because of the large DF_b / DF_a value at the β equal to 0.6, it is prudent that the required Design Factor, DF, be conservatively based on DF_b; that is,

$$DF = DF_b \qquad \text{(Eq. C2-12)}$$

In addition, basing the DF value on DF_b minimizes the effect of β on the Design Factor. As shown in Table C2-1, the probability ratios, R_P, corresponding to SDC 3, 4, and 5 are 4, 10, and 10, respectively. Slope ratios, A_R, range from a high of about 5.0 for some central and eastern U.S. hazard curves at $H_D = 4 \times 10^{-4}$ and 1-Hz frequency to a low of about 1.75 for California sites at $H_D = 1 \times 10^{-4}$.

Tables C2-2 and C2-3 present the resulting Design Factors, DF, corresponding to R_P ratios of 4.0 and 10.0, respectively, over a range of A_R values from 1.5 to 6.0 for β values of 0.3, 0.4, 0.5, and 0.6.

TABLE C2-2. Design Factors, DF, Corresponding to Probability Ratio, R_P, of 4.0 (SDC-3 Case)

A_R	β 0.3	β 0.4	β 0.5	β 0.6	Approx. Values
1.5	0.748	0.803	0.912	1.096	0.8
1.75	0.765	0.777	0.823	0.907	0.8
2	0.8	0.79	0.807	0.853	0.8
2.25	0.84	0.816	0.816	0.839	0.83
2.5	0.882	0.847	0.835	0.843	0.87
2.75	0.924	0.881	0.858	0.856	0.9
3	0.966	0.915	0.884	0.873	0.93
3.25	1.007	0.949	0.912	0.893	0.96
3.5	1.048	0.983	0.939	0.914	0.99
3.75	1.088	1.017	0.968	0.937	1.02
4	1.127	1.05	0.996	0.96	1.04
4.25	1.165	1.083	1.024	0.983	1.07
4.5	1.202	1.116	1.052	1.006	1.1
4.75	1.239	1.148	1.079	1.03	1.12
5	1.275	1.18	1.107	1.053	1.14
5.25	1.311	1.211	1.134	1.076	1.16
5.5	1.346	1.241	1.16	1.099	1.19
5.75	1.38	1.271	1.187	1.122	1.21
6	1.414	1.301	1.213	1.145	1.23

TABLE C2-3. Design Factors, *DF*, Corresponding to Probability Ratio, R_P, of 10.0 (SDC-4 and SDC-5 Cases)

A_R	β 0.3	β 0.4	β 0.5	β 0.6	Approx. Values
1.5	0.879	0.943	1.071	1.288	1
1.75	0.956	0.971	1.028	1.134	1
2	1.054	1.041	1.064	1.123	1.04
2.25	1.16	1.127	1.127	1.159	1.15
2.5	1.27	1.22	1.202	1.214	1.25
2.75	1.383	1.317	1.284	1.28	1.35
3	1.496	1.416	1.369	1.351	1.44
3.25	1.161	1.517	1.457	1.427	1.54
3.5	1.725	1.619	1.547	1.505	1.63
3.75	1.841	1.721	1.637	1.585	1.73
4	1.956	1.824	1.729	1.666	1.82
4.25	2.072	1.927	1.821	1.748	1.91
4.5	2.188	2.031	1.914	1.831	2
4.75	2.304	2.134	2.077	1.915	2.09
5	2.42	2.238	2.1	1.998	2.17
5.25	2.536	2.342	2.193	2.082	2.26
5.5	2.652	2.446	2.287	2.167	2.35
5.75	2.769	2.55	2.38	2.251	2.43
6	2.885	2.655	2.474	2.336	2.52

The Design Factors, *DF*, shown in Tables C2-2 and C2-3 can be reasonably approximated by

$$DF = \text{Maximum}\ (DF_1, DF_2) \quad \text{[Eq. C2-13(a)]}$$

$$DF_2 = 0.6A_R^{\alpha} \quad \text{[Eq. C2-13(b)]}$$

where DF_1 and α are defined in Table C2-4. The approximate *DF* values from Eqs. [C2-13(a)] and [C2-13(b)] are also shown in Tables C2-2 and C2-3. The approximate values are much better in the central region of A_R ratios of 2.0 through 4.5 because the sensitivity of *DF* to β is much less in this central region.

C2.2.1.3 Demonstration that the Design Factor Approach Reasonably Achieves Target Performance Frequencies, P_F

For this demonstration, the median of the nine EUS hazard curves and the California hazard curve shown in Fig. C2-1 at both 10 Hz and 1 Hz will be used. Normalized spectral acceleration values, *SA*, versus annual exceedance frequencies, $H_{(SA)}$, for the four hazard curves considered, are shown in Table C2-5.

TABLE C2-4. Design Factor Parameters

R_P	SDC	DF_1	α
4	3	0.80	0.40
10	4 and 5	1.00	0.80

For each of these four hazard curves and for SDC 3, 4, and 5, each of the following is tabulated in Table C2-6:

a. Target probabilistic performance goal
b. UHRS spectral acceleration, SA_{UHRS}, defined at the appropriate hazard exceedance frequency given in Table C2-1
c. Slope ratio, A_R, computed from Eq. (C2-3)
d. Design Factor, *DF*, computed from Eqs. [C2-13(a)] and [C2-13(b)] using the appropriate Design Factor parameters from Table C2-4
e. Resulting DBE spectral acceleration from Eq. (C2-1)
f. Resulting 10% conditional probability of failure capacity, $C_{10\%}$, from

$$C_{10\%} = 1.5\ SA_{DBE} \quad \text{(Eq. C2-14)}$$

corresponding to the nominal factor of safety being defined by Eq. [C2-10(b)]
g. Resulting actually achieved unacceptable performance probabilities, P_F, obtained by numerical integration of Eq. (C2-6) using the actual seismic hazard curves defined in Table C2-5 instead of the approximation defined by Eq. (C2-4) for $\beta = 0.3$, 0.4, 0.5, and 0.6 cases.

TABLE C2-5. Typical Normalized Spectral Acceleration Hazard Curve Values

	Eastern U.S.		California	
$H_{(SA)}$	1 Hz *SA*	10 Hz *SA*	1 Hz *SA*	10 Hz *SA*
5×10^{-2}	0.014	0.018	0.087	0.046
2×10^{-2}	0.027	0.034	0.13	0.072
1×10^{-2}	0.045	0.055	0.175	0.100
5×10^{-3}	0.07	0.089	0.236	0.139
2×10^{-3}	0.143	0.169	0.351	0.215
1×10^{-3}	0.235	0.275	0.474	0.334
5×10^{-4}	0.383	0.424	0.629	0.511
2×10^{-4}	0.681	0.709	0.814	0.762
1×10^{-4}	1.00	1.0	1.0	1.0
5×10^{-5}	1.46	1.41	1.23	1.22
2×10^{-5}	2.35	2.13	1.61	1.51
1×10^{-5}	3.27	2.88	1.89	1.76
5×10^{-6}	4.38	3.65	2.2	2.05
2×10^{-6}	6.44	4.62	2.68	2.42
1×10^{-6}	8.59	5.43	3.1	2.72
5×10^{-7}	10.34	6.38	3.58	3.06
2×10^{-7}	13.21	7.9	4.24	3.56
1×10^{-7}	15.9	9.28	4.67	3.84

The achieved performance probabilities, P_F, shown in Table C2-6, are based on the seismic acceptance criteria satisfying Eq. [C2-10(b)]. Correspondingly, Table C2-7 presents the achieved performance probabilities, P_F, when the seismic acceptance criteria satisfy Eq. [C2-10(a)]. If the acceptance criteria satisfy both Eqs. [C2-10(a)] and [C2-10(b)], then the achieved performance probabilities, P_F, are the lesser of those shown in Tables C2-6 and C2-7.

If the seismic acceptance criteria only satisfy Eq. [C2-10(b)], then Table C2-6 shows that the ratio of achieved to target performance probabilities range from 1.32 to 0.81, with a mean ratio of 0.97 for the cases studied. If the seismic acceptance criteria only satisfy Eq. [C2-10(a)], then Table C2-7 shows that the ratio of achieved to target performance probabilities ranges from 1.47 to 0.46, with a mean ratio of 0.88 for the cases studied. Only satisfying Eq. [C2-10(a)] results in considerably larger scatter in the achieved performance probabilities than does satisfying Eq. [C2-10(b)]. However, satisfying either Eq. [C2-10(a)] or [C2-10(b)] produces on average a conservative bias to the achieved performance probabilities, and these probabilities are never unconservative by more than a factor of about 1.5. If, as expected, the seismic acceptance criteria satisfy both Eq. [C2-10(a)] and [C2-10(b)], the ratio of the achieved to target performance probabilities as shown in Table C2-8 range from 1.17 to 0.46, with a mean ratio of 0.80. In this case, the achieved performance probabilities are never unconservative by more than a factor of about 1.2.

The specified *DF* values coupled with the dual acceptance criteria specified achieve a P_F no more than 1.17 (unconservative) times the target. The penalty is that the achieved P_F might be as low as 0.46 (conservative) times the target. This variability in achieved P_F cannot be avoided for any simple criteria because P_F varies by about a factor of two as a function of β. The goal has been to specify *DF* values that accurately

TABLE C2-6. Comparison of Actually Achieved Performance Probabilities, P_F, Versus Target Goals for Various Cases
Based on Seismic Acceptance Criteria Satisfying Eq. [C2-10(b)]

								Achieved Performance Probabilities P_F ($\times 10^{-5}$)			
SDC	Target Prob. Performance Goals ($\times 10^{-5}$)	Hazard Curve	UHRS	A_R	*DF*	*DBE DRS*	$C_{10\%}$	$\beta =$ 0.30	$\beta =$ 0.40	$\beta =$ 0.50	$\beta =$ 0.60
3	10.0	EUS 1 Hz	0.441	3.72	1.01	0.45	0.67	11.7	10.4	9.5	8.5
		EUS 10 Hz	0.431	3.25	0.96	0.46	0.69	11.4	10.0	9.1	8.6
		Calif. 1 Hz	0.670	1.96	0.80	0.54	0.81	9.5	8.8	8.7	9.3
		Calif. 10 Hz	0.563	2.28	0.83	0.47	0.70	11.3	9.8	9.0	8.7
4	4.0	EUS 1 Hz	0.441	3.72	1.72	0.76	1.14	4.5	4.0	3.6	3.4
		EUS 10 Hz	0.431	3.25	1.54	0.74	1.11	4.4	3.9	3.6	3.4
		Calif. 1 Hz	0.670	1.96	1.03	0.69	1.03	4.0	3.8	4.0	4.5
		Calif. 10 Hz	0.563	2.28	1.16	0.65	0.98	4.3	3.8	3.6	3.7
5	1.0	EUS 1 Hz	1.00	3.27	1.55	1.55	2.32	1.09	0.96	0.88	0.84
		EUS 10 Hz	1.00	2.88	1.40	1.40	2.10	1.03	0.90	0.83	0.81
		Calif. 1 Hz	1.00	1.89	1.00	1.00	1.50	1.04	1.02	1.11	1.32
		Calif. 10 Hz	1.00	1.76	1.00	1.00	1.50	0.84	0.83	0.89	1.02

TABLE C2-7. Comparison of Actually Achieved Performance Probabilities, P_F, Versus Target Goals for Various Cases
Based on Seismic Acceptance Criteria Satisfying Eq. [C2-10(a)]

SDC	Target Prob. Performance Goals ($\times 10^{-5}$)	Hazard Curve	$C_{1\%}$	Achieved Performance Probabilities P_F ($\times 10^{-5}$)			
				$\beta =$ 0.30	$\beta =$ 0.40	$\beta =$ 0.50	$\beta =$ 0.60
3	10.0	EUS 1 Hz	0.45	14.2	10.4	7.9	6.1
		EUS 10 Hz	0.46	13.8	10.0	7.4	5.6
		Calif. 1 Hz	0.54	12.8	8.5	6.2	4.9
		Calif. 10 Hz	0.47	14.7	9.6	6.7	4.9
4	4.0	EUS 1 Hz	0.76	5.4	3.9	2.9	2.3
		EUS 10 Hz	0.74	5.3	3.8	2.8	2.1
		Calif. 1 Hz	0.69	5.6	3.7	2.7	2.2
		Calif. 10 Hz	0.65	5.7	3.6	2.6	2.0
5	1.0	EUS 1 Hz	1.55	1.33	0.93	0.69	0.52
		EUS 10 Hz	1.40	1.30	0.87	0.62	0.46
		Calif. 1 Hz	1.00	1.47	0.96	0.73	0.61
		Calif. 10 Hz	1.00	1.22	0.78	0.58	0.48

achieve the Target Performance Goal for low variability failure modes (β between 0.3 and 0.4) while accepting increased conservatism for larger variability failure modes (β larger than 0.4).

C2.2.2 Vertical Ground Motion

In many cases, the modeling of horizontal ground motions at various depths in the soil column can be made with reasonable confidence using the simplifying assumption of upward traveling shear waves. The current state of the art for computation of vertical ground motions using analytic modeling alone cannot be made with the same degree of confidence. Therefore, vertical DRS are often derived by scaling the corresponding horizontal design spectrum, with frequency-dependent scale factors often selected from the empirical database.

Where the dominant earthquakes contributing to the seismic hazard are relatively distant, the vertical DRS can be obtained by scaling the ordinates of the horizontal spectrum by two-thirds throughout the entire frequency range. Where near-field earthquakes are significant contributors to the hazard, the ratio of vertical to horizontal spectral ordinates can be taken as at least unity for frequencies above 5 Hz, 2/3 at frequencies below 3 Hz, and a transition from 2/3 to 1 for frequencies between 3 Hz and 5 Hz, unless site-specific studies are conducted.

TABLE C2-8. Ratios of Achieved to Target Performance Probabilities

SDC	Hazard Curve	Ratio of Achieved to Target P_F			
		$\beta = .3$	.4	.5	.6
3	EUS 1 Hz	1.17	1.04	0.79	0.61
	EUS 10 Hz	1.14	1.00	0.74	0.56
	Calif. 1 Hz	0.95	0.85	0.62	0.49
	Calif. 10 Hz	1.13	0.96	0.67	0.49
4	EUS 1 Hz	1.12	0.98	0.72	0.58
	EUS 10 Hz	1.10	0.95	0.70	0.52
	Calif. 1 Hz	1.00	0.92	0.68	0.55
	Calif. 10 Hz	1.08	0.90	0.65	0.50
5	EUS 1 Hz	1.09	0.93	0.69	0.52
	EUS 10 Hz	1.03	0.87	0.62	0.46
	Calif. 1 Hz	1.04	0.96	0.73	0.61
	Calif. 10 Hz	0.84	0.78	0.58	0.48

It has been assumed in this context that the transition source-to-site distance between near- and far-field events can be selected as 15 km. Recent studies, however, have shown that ratios exceeding 2/3 may extend to even greater distances. In addition, recent observations indicate that the commonly adopted ratio of vertical to horizontal response spectra of 2/3 may be significantly exceeded over a wide range of frequencies of interest (Refs. [C2-3] and [C2-4]). The empirical database indicates that this ratio can be exceeded for distances as far as 40 km from the source, particularly at soil sites. Results from numerical modeling studies indicate that these ratios also increase to higher frequencies as rock velocities increase for eastern North America (ENA) sites. Therefore, it is recommended that where vertical motions are important contributors to development of seismic loads, site-specific evaluations be conducted to quantify these scale factors.

SECTION C2.3 METHOD TO DEFINE THE DESIGN RESPONSE SPECTRA AT VARIOUS DEPTHS IN THE SITE PROFILE

The current recommendation presents a method of convolving the ground motion associated with each of two bedrock outcrop UHRS spectra, at two return periods a factor of 10 apart in annual probability of exceedance (APE), to the ground surface or any other depth in the soil column. It is recommended that the convolution at any one APE consist of a set of convolution calculations that adequately capture the effects of potential variability and uncertainty in properties of the soils of the site profile. The mean of the spectra (computed at the free ground surface or other depth of interest) resulting from this set of computations is then used as the spectrum associated with the given APE. Enough sets of convolution calculations need to be performed to provide a stable estimate of the mean spectrum at the depth of interest. Sets of 30 or more site response computations have often been used for such estimates, in which a "Monte Carlo" approach is used to randomly select properties of individual soil layers.

In performing such calculations, information is required on low strain shear wave velocity, V_S, layer thickness, strain dependent shear modulus reduction, and hysteretic damping properties for each soil layer of the site profile, as well as its total depth to bedrock. Uncertainty in these parameters is also required to define the range in values that needs to be captured in the calculations. It is typical to use a coefficient of variation (COV) of unity in the estimate of V_S when good estimates of the variability in this parameter are unavailable for any soil layer. The recommendation in ASCE 4-98 indicates that a minimum value of COV of 0.5 should be used for this parameter in determining estimates of mean surface response spectra.

In performing the site response calculations, it is recommended that the seismic hazard be deaggregated to determine mean magnitudes and distances associated with frequencies of 1 Hz and 10 Hz. These M and D values are used to generate spectral shapes which are then scaled back to the UHRS at 1 Hz and 10 Hz, respectively. If the envelope spectrum from these two M and D characteristic events does not fall below the UHRS by more than 10% at any one frequency, the site amplification factors can then be generated from sets of convolution calculations generated from these two controlling events. If the envelope spectrum does fall more than 10% below the UHRS at any one frequency, a third characteristic event may need to be incorporated, using a spectral shape scaled to the UHRS at the intermediate frequency. This spectral shape should be appropriate for the characteristic event appropriate for the hazard at the intermediate frequency. If the spectra associated with these controlling events do not differ significantly, a single set of convolutions may be performed using a single broad-banded spectrum enveloping the UHRS and the two (or more) scaled spectra. Examples of scaling appropriate spectral shapes to the UHRS are presented in Refs. [C2-5] and [C2-6].

It should be noted that the selection of characteristic events associated with the UHRS at bounding frequencies of 1 Hz and 10 Hz is considered appropriate for relatively stiff structures typical of reactor systems. For softer structures or for structures situated on or in soft soil sites, the appropriate frequencies to consider as bounding frequencies may be modified somewhat to ensure that an appropriate range of potentially damaging events is considered.

Caution needs to be exercised when performing such sets of convolution calculations using the bedrock UHRS or a broad-banded envelope spectrum, particularly for deeper or softer soil sites. This involves a concern with the possibility of overdriving the soil column when using synthetic ground motions whose spectrum significantly exceeds the target bedrock outcrop spectrum associated with the characteristic events. By using motions that are more energetic than needed to envelop the targets, calculated soil strains may become unnecessarily large, leading to potentially significant increases in computed soil damping and modulus reduction. For deeper sites, this overstraining can lead to shifts in soil column frequencies, which can then influence computed site amplification factors.

After generating the mean response spectrum at the free ground surface or other depth of interest in the site profile, the Design Factor, DF, is then obtained from the resulting surface spectra as described in Section 2.2. The DRS is then determined by applying the *DF* to the resulting surface spectrum at the APE of interest. The intent of the evaluation is to arrive at a risk-consistent DRS at the depth of interest that is consistent with the bedrock outcrop spectra and that can be used as input to the seismic analyses of (near) surface founded structures.

SECTION C2.4 CRITERIA FOR DEVELOPING SYNTHETIC OR MODIFIED RECORDED TIME HISTORIES

The objective of the development of a synthetic time history or multiple time histories is to generate accelerogram(s) that have response spectra that closely match or envelop a given target spectrum and that, at the same time, possesses characteristics which are appropriate for the tectonic environment. In addition, the accelerogram(s) must have adequate power at all frequencies considered important to site response, SSI, or structural response evaluations. As described in Section 2.3, the output from a seismic hazard evaluation typically used in these types of studies consists of a UHRS appropriate for the APE together with characteristic events (*M*/*D* pairs) controlling at given response frequencies. Details on how such information can be determined from the seismic hazard analysis are described in Ref. [C2-7].

Based on empirical data, information is available (Ref. [C2-5]) on typical properties of events that have been categorized into *M*/*D* bins, including the effects of site characteristics on these properties. These properties include ranges of strong motion duration, peak velocity (*V*), displacement (*D*) and acceleration (*A*), and ratios of *V*/*A* and AD/V^2. Since these parameters are considered important in structural and site evaluation problems, the synthetic motion developed to represent a given *M*/*D* pair should have similar properties as the empirical data in the appropriate *M*/*D* bins. In generating synthetic motions, one procedure typically used is to select an appropriate Fourier phase spectrum representing the motion and then modify the corresponding Fourier amplitude spectrum iteratively until the response spectrum of the motion closely envelops the target spectrum.

The phase spectrum can be randomly selected or selected from the available database of recorded events. It is generally preferred that phase spectra be selected from the available database, which then results in acceleration, velocity, and displacement time histories that "look" reasonable (i.e., they resemble actual earthquake recordings from the appropriate *M*/*D* bin). However, for many site and SSI response analyses that use linear or equivalent linear methods of analysis, the phase spectrum has been found to have little impact on computed site response or in-structure response spectra (Ref. [C2-8]). However, when trying to represent near-field events where unusual characteristics of the recorded motions have been noted, or when determining site or structural responses including strong nonlinear behavior, phasing may play a more significant role. Therefore, it is recommended that the generated synthetic motion have a phase spectrum selected from appropriate empirical data.

Spectral accelerations at 5% damping should be computed at a minimum of 100 points per frequency decade, uniformly spaced over the log frequency scale from 0.1 Hz to 50 Hz or the Nyquist frequency. This number of points eliminates concern with the accuracy of the computed spectrum and can easily be accommodated with current computer capability. If the target response spectrum is defined in the frequency range from 0.2 Hz to 25 Hz, the comparison of the synthetic motion response spectrum with the target spectrum should be made at each frequency computed in this frequency range.

The objective is to generate an accelerogram whose spectrum provides a mean-based fit to the target spectrum. Therefore, many more points can fall below the target spectrum than has previously been required (Ref. [C2-9]). In Ref. [C2-9], requirements are specified which allow no more than 5 points to fall below the target. Since the objective in this development is to generate motions whose spectra are not biased high or low with respect to the target, about as many points can fall below the target as can exceed the target. However, to prevent large frequency ranges where the spectrum falls below the target, no more than nine contiguous frequency points are allowed to fall below in one frequency band. This number of points corresponds to a frequency window $\pm 10\%$ centered on the frequency of interest, if the frequency points are selected as described above.

To ensure that no frequency gaps occur in the motion, the exceedance of the computed spectrum above the specified target spectrum is limited to no more than 30% of the target. The computed spectrum should still not fall below the target by more than 10%. If these general criteria are followed, the matching requirements to the 5% damped response spectrum should be adequate to ensure that no gaps in the PSD or Fourier amplitude spectrum will occur over a significant frequency range. There is then no special need to further evaluate the PSD of the ground motion. If the 5% damped response spectrum of the generated accelerogram exceeds the target spectrum by more than 30%, the PSD of the accelerogram needs to be generated and shown to have adequate power at all frequencies of interest in the analysis as required in ASCE 4-98.

Directional correlation coefficients between pairs of records to be used in site response and/or SSI analyses are typically required to be relatively low to ensure that the structure or structural element cannot be oriented in the analysis in such a manner as to minimize some important directional response quantity of interest. However, if the limiting value is made too low, a significant number of empirical recordings in any earthquake category may be unnecessarily eliminated from further consideration as a seed for generating design ground motions. Since the response quantity is a function of the structural characteristics and not of the empirical bin data sets, it is recommended that the limit for correlation coefficient between any two design ground motions be set at 0.3. For values less than this limit, it is felt that no significant reduction in directional response can be attained.

Nonlinear dynamic analyses of soil–structure systems require specification of appropriate acceleration records (accelerograms). Since design ground motion criteria are typically specified in terms of response spectra, it becomes necessary to select appropriate recorded or modified recorded accelerograms to use in the analyses, which need to be suitably scaled to the criteria spectrum (Refs. [C2-10] and [C2-11]). Accelerograms recorded at soil sites are to a large extent dominated by the local dynamic characteristics of the site. Since soil sites are highly variable, matching of soil sites where records have been obtained to a soil site under consideration is difficult, if not impossible. Therefore, in general, it is recommended that accelerograms representing appropriate bedrock outcrop motions be selected, and site-specific convolution through the soil column, as described in Section 2.3, be performed to incorporate site-specific amplification effects.

Since recordings are typically organized into *M*/*D* bins, the first step is to identify the appropriate *M*/*D* bins that impact the site DRS. Depending on the procedure used in determining the DRS, the dominant contributing events can be selected from either deaggregation of the seismic hazard at important frequencies of interest for the structural analysis, or from events appropriate at the displacement, velocity, and acceleration segments of the response spectrum. If a probabilistic seismic hazard analysis is used to define the DRS, these characteristic events can be selected as dominating the hazard at low frequencies (typically defined at 1 Hz or lower), at an intermediate frequency, and at high frequencies (typically defined at 10 Hz). Alternatively, for all identified credible events, the appropriate *M*/*D* bins can be selected which characterize the displacement, velocity, and acceleration segments of the response spectrum.

Once the characteristic *M*/*D* bins are selected, the individual recorded ground motion needs to be scaled such that its response spectrum matches the DRS over the frequency range of interest. The scaled accelerogram can then be used as input to the nonlinear soil-structural model and computed responses evaluated for acceptance.

REFERENCES FOR SECTION C2.0

[C2-1] Kennedy, R.P., and Short, S.A. (1994). "Basis for seismic provisions of DOE-STD-1020." *UCRL-CR-111478*, U.S. Dept. of Energy, Washington, D.C.

[C2-2] ANS and Independent Plant Evaluation for External Events (1983). "PRA procedures guide." *NUREG/CR-2300*, Vol. 2, Chap. 10. Prepared for the U.S. Nuclear Regulatory Commission, Washington, D.C.

[C2-3] Abrahamson, N.A., and Silva, W.J. (1997). "Empirical response spectral attenuation relations for shallow crustal earthquakes." *Seismological Research Letters*, 68(1).

[C2-4] Silva, W.J. (1997). "Characteristics of vertical strong motions for applications to engineering design." *Proc., FHWA/NCEER Workshop on the National Representation of Seismic Ground Motion for New and Existing Highway Facilities, Tech. Rpt. NCEER-97-0010*. National Center for Earthquake Engineering Research, Buffalo, N.Y.

[C2-5] McGuire, R.K., Silva, W.J., and Costantino, C.J. (2001). "Technical basis for revision of regulatory guidance on design ground motions Hazard- and risk-consistent ground motion spectra guidelines." *NUREG/CR-6728*, U.S. Nuclear Regulatory Commission, Washington, D.C.

[C2-6] McGuire, R.K., Silva, W.J., and Costantino C.J. (2002). "Technical basis for revision of regulatory guidance on design ground motions: Hazard- and risk-consistent ground motion spectra for two sites." *NUREG/CR-6769*, U.S. Nuclear Regulatory Commission, Washington, D.C.

[C2-7] U.S. Nuclear Regulatory Commission (1997). "Identification and characterization of seismic sources and determination of safe shutdown earthquake ground motion." *Regulatory Guide 1.165*, Washington, D.C.

[C2-8] Costantino, C.J. (2001). "A study on sensitivity of site and SSI response to phase characteristics of earthquake ground motions." Report prepared for Savannah River Operations Office, Dept. of Energy, Aiken, S.C.

[C2-9] U.S. Nuclear Regulatory Commission (1996). "U.S. NRC standard review plan." *NUREG-0800*, Rev. 2, Washington, D.C.

[C2-10] Shome, N., Cornell, C.A., Bazzurro, P., and Carballo, J.E. (1998). "Earthquakes, records and nonlinear responses." *Earthquake Spectra*, Aug.

[C2-11] Carballo, J.E., and Cornell, C.A. (1998). "Input to nonlinear structural analysis: Modification of available accelerograms for different source and site characteristics." *Proc., 6th U.S. National Conf. on Earthquake Engineering*, Earthquake Engineering Research Institute, Oakland, Calif.

SECTION C3.0 EVALUATION OF SEISMIC DEMAND

SECTION C3.3 NONLINEAR ANALYSIS

Based on relatively limited numerical studies using a structural model that primarily responds in its fundamental mode (Ref. [C3-1]), it has been shown that the parameter of primary importance in estimating median values of nonlinear response measures is the spectral acceleration at the fundamental period of the structure. It is also indicated that the dispersion of the nonlinear response measure can be somewhat reduced if "local spectral averaging" is used to match the spectrum of the input accelerogram to the target spectrum. That is, the matching is done over a range of frequencies that is not too wide and is symmetric about the fundamental frequency.

These results indicate that obtaining a median estimate of nonlinear response within approximately, say, 20% with 95% confidence requires the use of about 6 to 7 independent accelerograms (all scaled to the target spectral acceleration). However, many more calculations would be required if an estimate of the 84% response were required.

SECTION C3.4 MODELING AND INPUT PARAMETERS

C3.4.1 Effective Stiffness of Reinforced Concrete Members

Table 3-1 is based on data from Table 6-5 of FEMA 356 [Ref. C3-2]. For additional information on effective stiffness, consult Section C6.4.1.2 of FEMA 274 [Ref. C3-3], and consult ASCE 4.

C3.4.3 Damping Values for SSCs

Cable tray and conduit damping values are based on extensive testing (Ref. [C3-3]). These tests clearly demonstrated that a substantial amount of energy is absorbed by friction between the cables and through bouncing of the cables (Ref. [C3-4]). This phenomenon was also observed to be amplitude dependent (i.e., the greater the input level, the more pronounced were the losses). Equating the losses to an equivalent viscous damping resulted in predicted equivalent viscous damping of up to 50% in some cases. After plotting the results of the several hundred earthquake type vibration tests and cable tray systems, a conservative lower bound curve representing equivalent viscous damping as a function of input floor spectrum ZPA was plotted as shown in Figure C3-1.

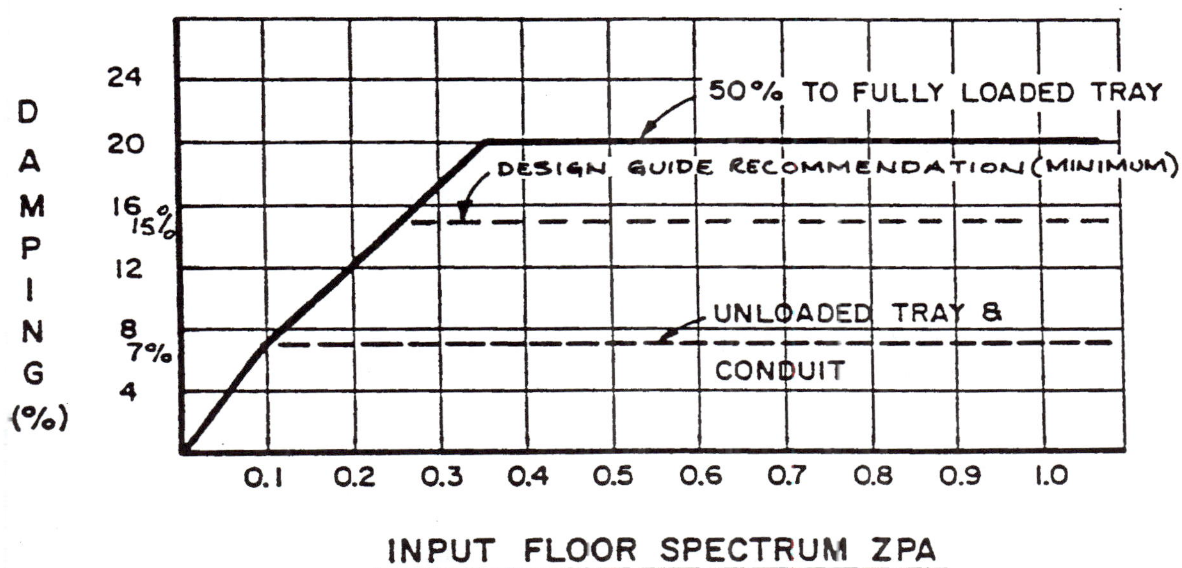

C3-1. Cable Tray Damping as a Function of Input ZPA

Based on Figure C3-1, the level of damping to be used in the analysis of cable tray support systems is a minimum of 15% of critical damping, for systems that are at least 50% full and with an in-structure response spectrum ZPA of 0.25 g. For the lightly loaded cable tray systems, systems with lower floor ZPAs or the systems with rigid fireproofing, the damping should be reduced as given in Table 3-2. Likewise, for raceway systems carrying mainly conduits, the lower damping values are applicable.

REFERENCES FOR SECTION C3.0

[C3-1] Shome, N., Cornell, C.A., Bazzurro, P., and Carballo, J.E. (1998). "Earthquakes, records and nonlinear responses." *Earthquake Spectra*, Aug.

[C3-2] FEMA. (2000). "Prestandard and commentary for the seismic rehabilitation of buildings." *FEMA 356*, Washington, D.C.

[C3-3] FEMA. (1997). "NEHRP commentary on the guidelines for the seismic rehabilitation of buildings." *FEMA 274*, Washington, D.C.

[C3-4] ANCO Engineers, Inc. (1978). "Cable tray and conduit raceway seismic test program: Release 4." *Report 1053-21.1-4*, Boulder, Colo.

SECTION C4.0 EVALUATION OF STRUCTURAL CAPACITY

SECTION C4.2 STRUCTURAL CAPACITIES

Capacity of structures may be based on a number of building codes. For nuclear facilities, the following codes are applicable:

ACI 349	For capacity calculations of reinforced concrete members.
AISC LRFD	For capacity calculations of steel members using the strength approach.
AISC N690	For allowable stress capacity calculations of steel members. Alternatively, capacities based on the allowable stresses may be multiplied by an appropriate factor to determine the design strength.
ACI 530	For determining the capacities of reinforced concrete masonry elements using the working stress approach.
IBC 2003	For determining the strength of reinforced masonry members using the strength design method.

In the strength design approach, the nominal capacity of a member is determined by following the procedures given in the applicable code. These capacities may be based on simple section properties and the material strength (e.g., cross-sectional area times the yield strength for tensile capacity) or on an empirical equation (e.g., in-plane shear capacity of reinforced concrete shear walls). Once the nominal capacity is determined, the code capacity is established by applying a capacity (strength) reduction factor. The code capacity is then compared with the demand.

In the allowable (working) stress approach, the design capacity is based on the elastic section properties and allowable stresses. For SDC-3 and higher category structures, capacity reduction factors and response modification factors are not applicable.

C4.2.3 Capacity of Low-Rise Concrete Shear Walls

Studies have shown that the shear strength of low rise concrete shear walls is overly conservative when predicted by recent versions of the ACI 318 code provisions (Refs. [C4-1] to [C4-3]), as well as ACI 349 provisions. This is particularly true for walls with height-to-length ratios of about 2.0 or less. Currently, special ACI code provisions for seismic design of structural concrete walls limit the strength contribution from concrete to a maximum of $3(f'_c)^{1/2}$ for low-rise shear walls. Barda (Ref. [C4-6]) determined that the ultimate shear strength of low-rise ($h_w/l_w < 1.0$) walls tested for the diagonal shear failure mode could be represented by the following relationship:

$$v_u = v_c + v_s$$

$$v_u = 8.3\sqrt{f'_c} - 3.4\sqrt{f'_c}\left(\frac{h_w}{l_w} - 0.5\right) + \rho_v f_y \qquad \text{(Eq. C4-1)}$$

where

v_u = Ultimate shear strength (psi)
v_c = Contribution from concrete (psi)
v_s = Contribution from steel reinforcement (psi)
f'_c = Concrete compressive strength (psi)
h_w = Wall height (in.)
l_w = Wall length (in.)
ρ_v = Vertical steel reinforcement ratio
f_y = Steel yield strength (psi)

A study conducted for the Diablo Canyon Nuclear Power Plant of the test data provided by Barda and other researchers found that the contribution from concrete in Eq. (C4-1) works well for wall aspect ratios up to 2.0 (Ref. [C4-7]). The contribution of concrete to the ultimate shear strength of the wall as a function of

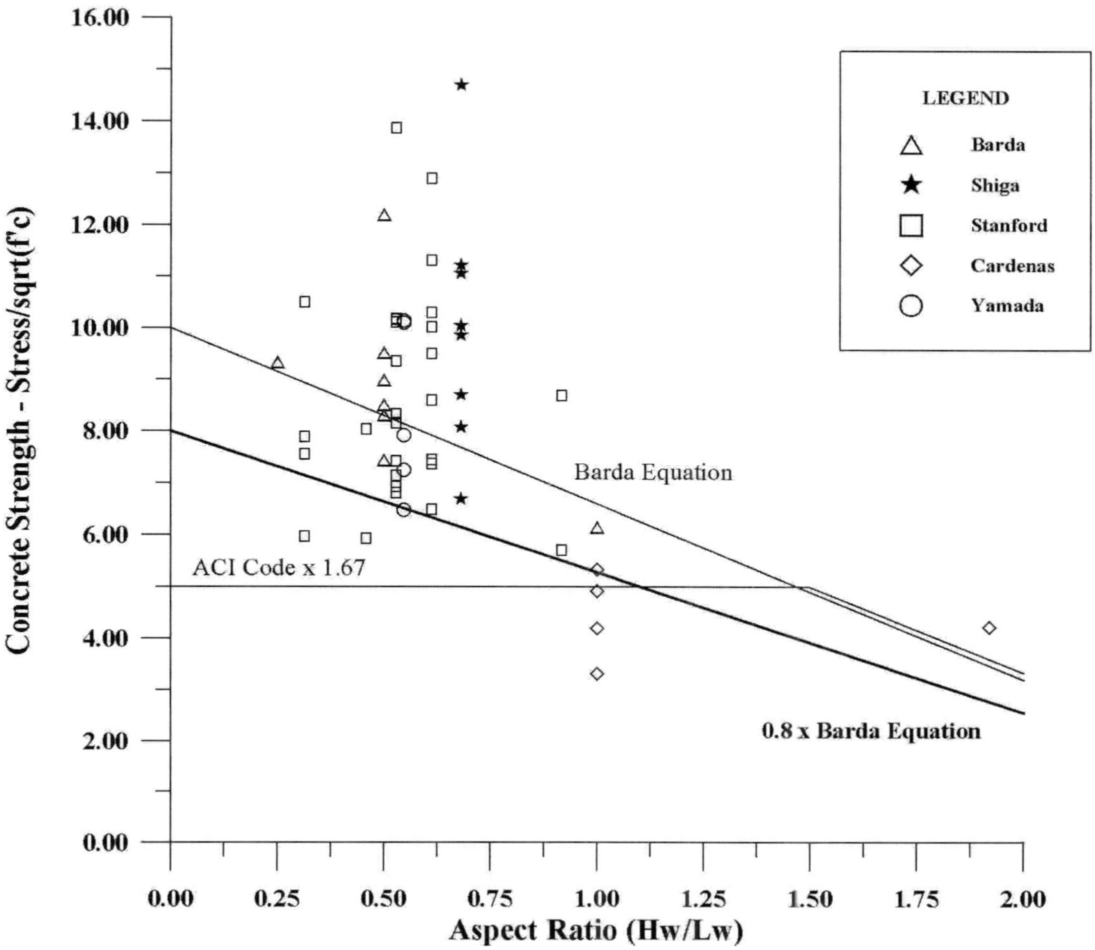

C4-1. Strength of Shear Walls

h_w/l_w is shown in Figure C4-1 (referred to as the "Barda Equation"), along with a line that represents 0.8 times the Barda Equation (i.e., Eq. (C4-4) for concrete contribution). Also shown in Figure C4-1 are the available test values (Refs. [C4-6] to [C4-12]) and the corresponding ACI 349 formulation. The part of the test data capacities due to reinforcing steel has been removed so that only the concrete capacity is reflected in the data points. Also, the ACI code curve has been increased by a factor of 1.67 to make it comparable to the Barda Equation. This corresponds to the case when the default value of d equal to 0.6 l_w is used with the Barda Equation, and effectively l_w is equal to d in the ACI approach. The tests included load reversals and varying reinforcement ratios and h_w/l_w ratios. Comparing the Barda Equation to 1.67 times the ACI code equation, it is seen that the ACI code severely underestimates the capacity of concrete at low aspect ratios.

Testing was performed with and without axial loads, and an increase in shear capacity of $N_A/4l_wt_n$ was recommended, where N_A is the axial load in pounds and t_n is the wall thickness in inches. In order to estimate the effects that the horizontal and vertical steel have, the steel contribution to wall shear strength was determined from test values for the range of $0.3 < h_w/l_w < 2$ (Ref. [C4-7]). Test data from Ref. [C4-6] and Refs. [C4-8] through [C4-12] were used to develop the following equation for the effective steel shear strength:

$$v_{se} = Av_{sv} + Bv_{sh} \qquad \text{(Eq. C4-2)}$$

where

A, B are constants

$v_{sv} = \rho_v f_y$ = vertical steel contribution to shear strength

$v_{sh} = \rho_u f_y$ = horizontal steel contribution of shear strength

ρ_v = Vertical steel reinforcement ratios

ρ_u = Horizontal steel reinforcement ratios

The constants A and B were then calculated assuming that the concrete contribution to the ultimate strength is taken from Eq. (C4-1). Based on the results of the evaluation in Ref. [C4-7], the constants A and B are approximated as

$h_w/l_w < 0.5$	$A = 1$	$B = 0$
$0.5 < h_w/l_w < 1.5$	$A = -h_w/l_w + 1.5$	$B = h_w/l_w - 0.5$
$1.5 < h_w/l_w$	$A = 0$	$B = 1$

Figure C4-2 shows a plot of the A and B constants.

The median ultimate shear strength is given by

$$v_u = v_c + v_{se}$$

$$v_u = 8.3\sqrt{f'_c} - 3.4\sqrt{f'_c}\left(\frac{h_w}{l_w} - 0.5\right) + \frac{N_A}{4l_w t_n} + \rho_{se} f_y \quad \text{(Eq. C4-3)}$$

where $\rho_{se} = A\rho_v + B\rho_h$, with A and B determined as discussed above. Based on an evaluation of the same experimental data, the logarithmic standard deviation is estimated to be 0.20 (Ref. [C4-7]). Thus, a value for ultimate shear strength based on the minus one standard deviation is

$$v_u = \phi\left[8.3\sqrt{f'_c} - 3.4\sqrt{f'_c}\left(\frac{h_w}{l_w} - 0.5\right) + \frac{N_A}{4l_w t_n} + \rho_{se} f_y\right] \quad \text{(Eq. C4-4)}$$

The corresponding minus one standard deviation plot for the concrete strength is also shown in Figure C4-1 (0.8 times Barda Equation). A capacity reduction factor, $\phi = 0.8$, is recommended to be used with Eq. (C4-3) as reflected in Eq. (C4-4).

The data used to develop Eq. (C4-4) was from tests conducted on cantilever walls with lateral load

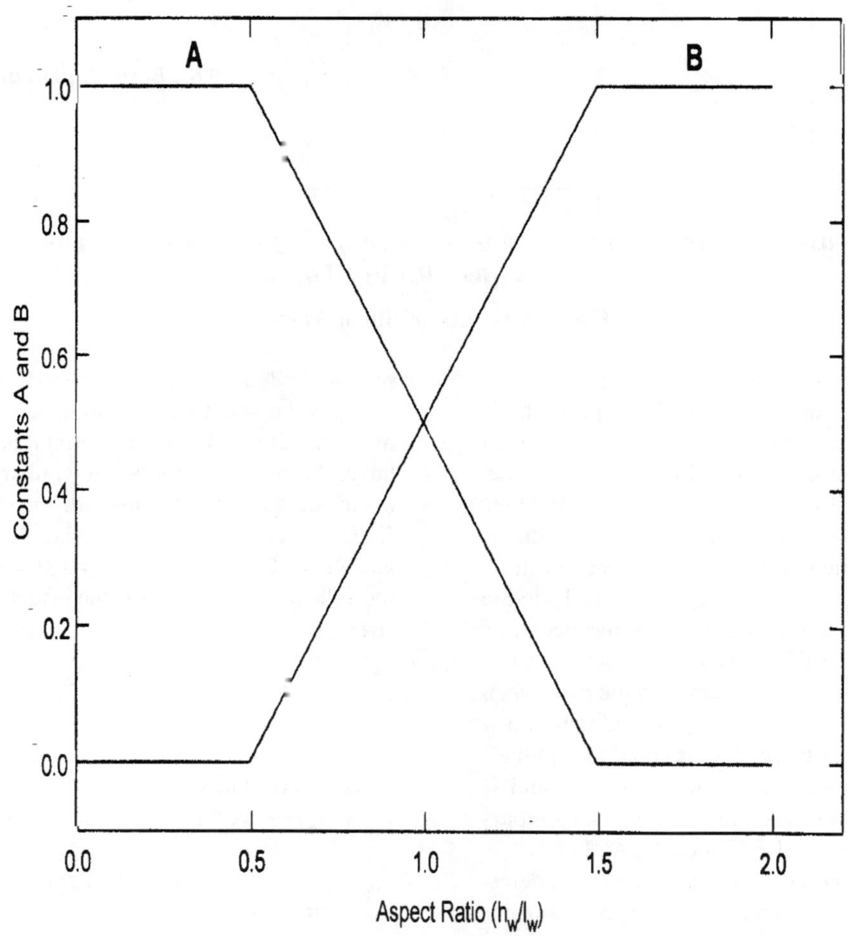

C4-2 Plots of Constants A and B

applied at the top of the wall, with constant reinforcement throughout the wall height. For multistory walls, the shear capacity should be checked both for the overall wall (multistories), in which case h_w is the total wall height, and for individual stories, in which case h_w is the individual story height. For multistory walls, it is not clear whether h_w should be the overall wall height or M/V, where M is the applied overturning moment at the base and V is the shear also at the base. It is more conservative to use h_w equal to the overall wall height. It will also be necessary to check multiple levels when the steel percentage varies up the wall height. When checking the overall shear wall, the area of horizontal reinforcement within intermediate floor slabs within a distance of four times the slab thickness from the wall may be included as part of the horizontal reinforcement of the wall. Because of limitations on the data from which Eq. (C4-4) was derived, ρ_{se} should not exceed 0.01 and the ultimate shear strength, v_u, should not exceed $20\phi(f_c')^{1/2}$ unless justification is provided.

As per ACI 349, the shear force demand should be that associated with a section located a distance $l_w/2$ or $h_w/2$, whichever is less, above the base of any wall segment being checked as long as the wall properties are not reduced between this level and the base.

The total shear capacity is

$$V_u = v_u \, d \, t_n \qquad \text{(Eq. C4-5)}$$

where d is the distance from the extreme compression fiber to the center of force of all reinforcement in tension, which may be determined from a strain compatibility analysis. If such an analysis is not performed, then

$$d = 0.6 \, l_w \qquad \text{(Eq. C4-6)}$$

REFERENCES FOR SECTION C4.0

[C4-1] ACI. (1971). "Building code requirements for reinforced concrete." *ACI 318-71*, Detroit.

[C4-2] ACI. (1989). "Building code requirements for reinforced concrete." *ACI 318-89*, Detroit.

[C4-3] ACI. (1995). "Building code requirements for reinforced concrete." *ACI 318-95*, Detroit.

[C4-4] ACI. (1999). "Building code requirements for masonry structures and commentary." *ACI 530-99/ASCE 5-99/TMS 402-99*, Detroit.

[C4-5] International Code Council, Inc. (2003). *International Building Code 2003*, Whittier, Calif.

[C4-6] Barda, F., Hanson, J.M., and Corley, W.G. (1976). "Shear strength of low-rise walls with boundary elements." *ACI Symposium on Reinforced Concrete Structures in Seismic Zones*, ACI, Detroit.

[C4-7] Benjamin, J.R., and Reed, J.W. (1983). *Recommended evaluation criteria for Diablo Canyon Nuclear Power Plant auxiliary building walls and diaphragms*. Jack R. Benjamin and Assoc., Inc., Mountain View, Calif.

[C4-8] T. Shiga, Shibata, A., and Tabahashi, J. (1973). "Experimental study on dynamic properties of reinforced concrete shear walls." *Proc., 5th World Conf. on Earthquake Engineering*, International Association for Earthquake Engineering, Rome.

[C4-9] Benjamin, J.R., and Williams, H.A. (1954). "Investigation of shear walls, part 6: Continued experimental and mathematical studies of reinforced concrete walled bents under static shear loading." Dept. of Civil Engineering, Stanford Univ.

[C4-10] Williams, H.A., and Benjamin, J.R. (1953). "Investigation of shear walls: Part 3, Experimental and mathematical studies of the behavior of plain and reinforced concrete walled bents under static shear loading." Dept. of Civil Engineering, Stanford Univ.

[C4-11] A. E. Cardenas, et al. (1973). "Design provisions for shear walls." *ACI Journal*, 70(3).

[C4-12 Oesterle, R.G., et al. (1979). "Earthquake resistant structural walls: Tests of isolated walls, Phase II." Construction Technology Laboratories (Division of PCA), Skokie, Ill.

SECTION C5.0 LOAD COMBINATIONS AND ACCEPTANCE CRITERIA FOR STRUCTURES

The seismic loading combination in this Standard combines a "best estimate" of the normal operating loads and the seismic load developed in Section 3.0. This Standard has acceptance criteria for both linear and nonlinear analyses of structures. A force-based acceptance criterion is utilized for linear analyses along with a check of the total elastic displacements. A displacement-based acceptance criterion is utilized for nonlinear analysis, with the provision that member capacities be limited to the product of the nominal material code capacity and the material code strength reduction factor, ϕ. Load combinations and acceptance criteria for systems and components are discussed in Section 8.0.

SECTION C5.1 LOAD COMBINATIONS

C5.1.1 General

The seismic loading combination in this document combines a "best estimate" of the normal operating loads and the seismic load developed in Section 3.0.

This seismic load combination is consistent with the seismic performance goals that form the basis of this criteria and is intended to supercede seismic load combinations given in material codes (e.g., ACI, AISC), model building codes (e.g., IBC), and other standards (e.g., ASCE 7). This Standard presumes that loading combinations for non-seismic loads are consistent with the material codes, model building codes, and ASCE 7.

C5.1.2.1 Seismic Loading Combinations for Strength-Based Acceptance Criteria

The seismic demand for elastic analyses is reduced by an inelastic force reduction factor, $F_{\mu S}$ to account for nonlinear energy absorption.

Effective Natural Frequency. Force reduction factors for rigid frames, F_μ, are highly dependent on frequency, as discussed in Section C.5.1.2.3; Ref. [C5-5] developed relationships for these force reduction factors at different frequencies. The force reduction factors in Table 5-1 are based on the assumption that rigid frames have natural frequencies less than about 3 Hz, which is typically at the frequency corresponding to the peak spectral acceleration.

For higher frequencies, the softening of the frame due to inelastic action can shift the frame's effective natural frequency to a region of higher spectral acceleration, as shown in Figure C5-1. In Ref. [C5-5], this effect is included in the frequency-dependent F_μ. In this Standard, a simplified approach is used to account for this effect by using the spectral acceleration of the effective natural frequency, f_e.

Given the idealized load deflection diagram in Figure C5-2, the elastic stiffness is K_e, while the secant stiffness of a nonlinear frame with ductility, μ, can be expressed as

$$K_{NL} = \frac{K_e \Delta y}{\mu \Delta y} = \frac{K_e}{\mu} \qquad \text{(Eq. C5-1)}$$

Assume that the effective frequency, f_e, is based on the nonlinear secant stiffness, K_{NL}. Note that this assumption leads to a conservative lower bound effective frequency. The ratio of the effective to elastic natural frequency, f_n, is

$$\frac{f_e}{f_n} = \frac{\sqrt{\dfrac{K_{NL}}{M}}}{\sqrt{\dfrac{K_e}{M}}} = \sqrt{\frac{1}{\mu}} \qquad \text{(Eq. C5-2)}$$

Since F_μ for rigid frames are based on about 3 Hz, which is typically in the constant velocity range, then $F_\mu \approx \sqrt{2\mu - 1}$, and a conservative estimate of the effective frequency is

$$f_e = \sqrt{\frac{2}{F_\mu^2 + 1}} f_n \qquad \text{(Eq. C5-3)}$$

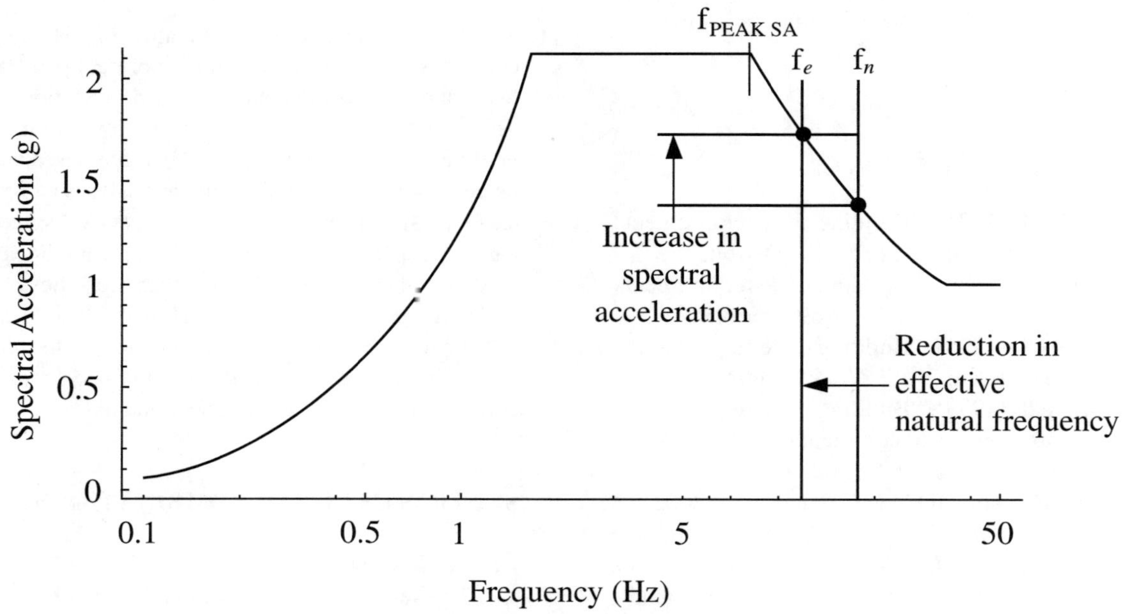

C5-1 Idealized Design Response Spectra

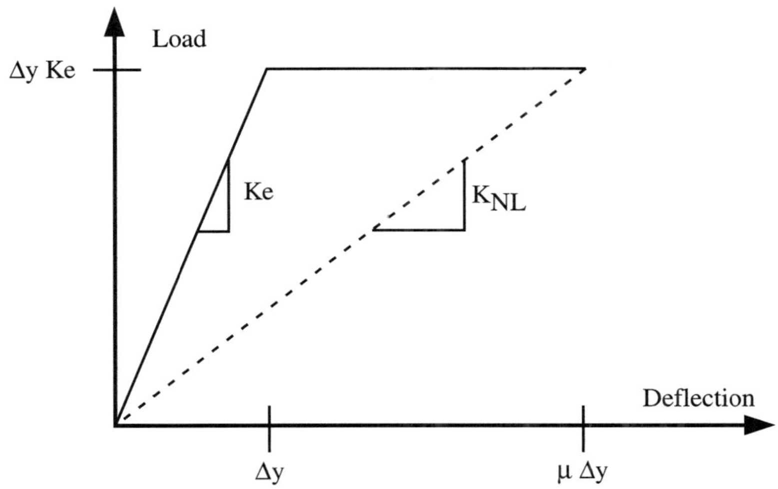

C5-2 Idealized Load Deflection Curve

The minimum value of the effective frequency is limited to $f_{PEAK\ SA}$ to force the use of the peak spectral acceleration for structures that soften through the peak of the spectra.

Note that this discussion is based on rigid frames. Structures with pinched hysteresis loops, such as low-rise shear walls and braced frames, have F_μ that do not vary significantly between 2 Hz and 8.5 Hz (Ref. [C5-11]).

Weak Story Effect. The F_μ modification for weak story effects is given by Eq. (5-3) (Ref. [C5-5]). This formulation is based on the assumption that all of the inelastic deformation is concentrated into a single weak story of an n-story frame. Eq. (5-3) was developed for displacement controlled structures where $F_\mu = \mu$ and is conservative for velocity and acceleration controlled structures. Numerical studies have also demonstrated that Eq. (5-3) is conservative for reasonable variations in mass and stiffness.

C5.1.2.2 Seismic Loading Combinations for Displacement-Based Acceptance Criteria

Nonlinear lateral displacements for an elastic analysis are approximated by $\mu\Delta_{elastic}$. Calculating the elastic lateral displacements with Eq. (5-1) and assuming that the force reduction factor, $F_{\mu S}$, is approximated by μ, then the nonlinear displacement is

$$\Delta_{NL} = \mu\Delta_{elastic} = \frac{\mu\left(1.0\dfrac{D_s}{F_{\mu s}}\right)}{\text{Stiffness}} \cong \frac{1.0D_s}{\text{Stiffness}} \quad \text{(Eq. C5-4)}$$

which is similar to the displacement calculated with Eq. (5-5). Thus, Eq. (5-5) is used to estimate the nonlinear displacements for an elastic analysis. Note that $F_{\mu S}$ is approximated by μ in the displacement portion of the spectra and is less than μ in the velocity and acceleration portions of the spectra. Thus, Δ_{NL} is larger than the elastic displacement with 1.0 D_s at medium to high frequencies. However, displacements in this frequency range are typically small and do not control the design. Thus, this unconservatism is acceptable.

C5.1.2.3 Inelastic Energy Absorption Factor, F_μ

The engineering community has known for some time that ductile structures are better at resisting seismic loads than are elastic structures. Building codes for conventional structures recognize the benefits of ductile construction and allow for design using a reduced seismic base shear to reflect the beneficial effects of ductile behavior. This reduced base shear is calculated using force reduction factors. However, building codes for conventional structures are primarily concerned with preventing the loss of human life and allow significant structural damage. This level of damage may not be appropriate for facilities containing nuclear materials. For example, it may be desirable to limit the seismic damage in a nuclear confinement structure to hairline cracks while a non–safety-related structure could suffer appreciable damage as long as it did not collapse on a safety-related structure. Thus, an appropriate set of design rules for a nuclear structure would allow for various force reduction factors dependent on the amount of damage that is acceptable. This acceptable damage level defines the Limit State.

A force reduction factor, R, that accounts for both the overstrength and the inelastic energy absorption of the structural system is used in the model building codes for conventional structures. Overstrength in a building system can be quantified using a pushover analysis or estimated from existing analyses of similar structures. Pushover analyses are not usually performed as part of the design process, and the use of existing analyses to estimate overstrength is problematic for nuclear facilities. Thus, the overstrength factor is conservatively omitted from the force reduction factors in Table 5-1. The overstrength factor in typical non-nuclear commercial building frames ranges from 2 to 3.5 and represents about one-half of the R factor used in building codes (ATC-2, 1974). An overstrength for a particular structure that is quantified by a pushover analysis may be included in the development of a facility specific force reduction factor, F_μ.

Inelastic energy absorption factors, F_μ, have been extensively studied. The element ductility, amount of strain hardening, shape of the hysteresis loops, building frequency, supporting soil type, and the influence of higher modes [single-degree-of-freedom versus multiple-degree-of-freedom (SDOF versus MDOF)] affect F_μ. The force reduction factors in this Standard are developed by the following four steps:

1. Selecting element ductilities, $\mu_{element}$, that are correlated to desired performance or Limit State.
2. Relating frequency dependence, secondary effects of strain hardening, hysteresis loop shape, and soil type to the force reduction factor for a SDOF system, $F_{\mu SDOF}$;
3. Develop relationships between the ductility of MDOF and SDOF systems; and
4. Combine steps 1, 2, and 3 to yield force reduction factors, F_μ, for design.

These four steps are described below and shown schematically in Figure C5-3.

First, acceptable damage for different Limit States is defined for each type of element. Four Limit States are used in this Standard that range from significant damage in Limit State A (LS-A) to essentially elastic behavior in LS-D. Intermediate Limit States consist of LS-C, which corresponds to immediate occupancy, and LS-B, which is defined as halfway between LS-A and LS-C. Story displacements or element rotations associated with each Limit State are selected based on the average element behavior, as shown in Figure C5-4. The summary of element data in Refs. [C5-6] and [C5-7] form the basis for the element ductilities and allowable deformations used in this Standard. The element is conservatively assumed to have an elasto-plastic behavior with a yield point at the nominal code capacity and an allowable inelastic deformation corresponding to the appropriate Limit State displacement. Element ductilities for each Limit State are derived on this assumption.

Second, the inelastic response of different SDOF systems to various ground motions has been extensively studied (Refs. [C5-8] to [C5-11]). Flexural structural systems are represented by hysteresis models having full stable hysteresis loops, while structural systems dominated by shear deformation or diagonal brace buckling are represented by pinched hysteresis models. Relationships developed by several researchers are examined and force reduction factors for a SDOF system, $F_{\mu SDOF}$, are determined considering structural system type, system frequency, and element ductility. These force reduction factors have a conservative bias when compared to literature relationships (Ref. [C5-5]).

Third, SDOF models neglect the contribution of higher modes and the relationship between component and global ductility. For some structures these factors may be important, especially at lower frequencies. Relationships between the response of MDOF systems and SDOF systems are presented in the literature (Ref. [C5-9] and Refs. [C5-12] to [C5-16]). The allowable element ductility, modified to account for MDOF effects, is used to develop frequency-dependent values of F_μ for elements used in each of the building types. The relationship between SDOF and MDOF F_μ for structures that have a soft or weak story (Ref. [C5-5]) is not included in the F_μ in Table 5-1 and is applied separately in this Standard.

Fourth, the resulting frequency-dependent F_μ values are functions of Limit State and natural frequency, as shown in Figure C5-3. Conservative frequency-independent F_μ factors for each type of framing system are selected based on building frequencies representative of that typically found in facilities containing nuclear materials. These frequency-independent F_μ factors are given in Table 5-1. The use of frequency-dependent F_μ factors, such as those given in Ref. [C5-5], is permissible.

Shear Walls. The F_μ given for shear walls are for in-plane bending moments and shears. An out-of-plane shear wall can be considered as a beam with a low axial load. Shear walls governed by out-of-plane shear are brittle unless special out-of-plane shear reinforcing has been installed. Brittle elements have $F_\mu = 1.0$.

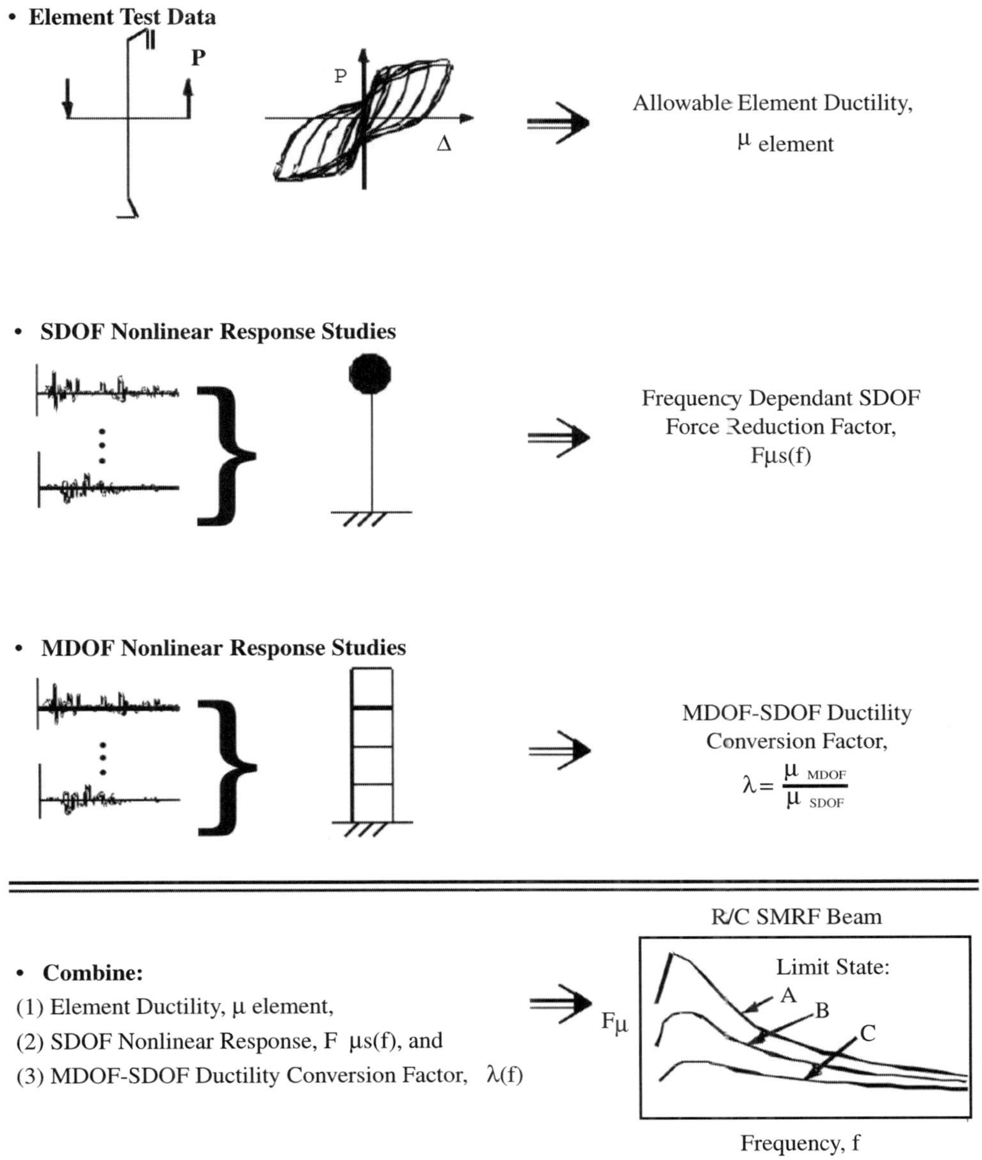

C5-3 $F\mu$ Development

Elements that May Experience Ratcheting. The F_μ developed in this Standard did not consider external forces that act in the same direction as the seismic load. If these forces are sufficiently large, they have the potential to prevent the structure from unloading and cause displacements that are significantly larger in one direction than the other. Elements with a potential for ratcheting include basement walls with at-rest lateral soil loads subject to dynamic lateral soil loads and slabs with gravity loads subject to vertical seismic motion.

The F_μ in Table 5-1 should not be used for structures with significant ratcheting potential. Alternatives include the following:

- Designing the ratcheting sensitive element with $F_\mu = 1.0$
- Developing F_μ that includes ratcheting
- Performing a nonlinear analysis that includes both the normal operating and seismic forces.

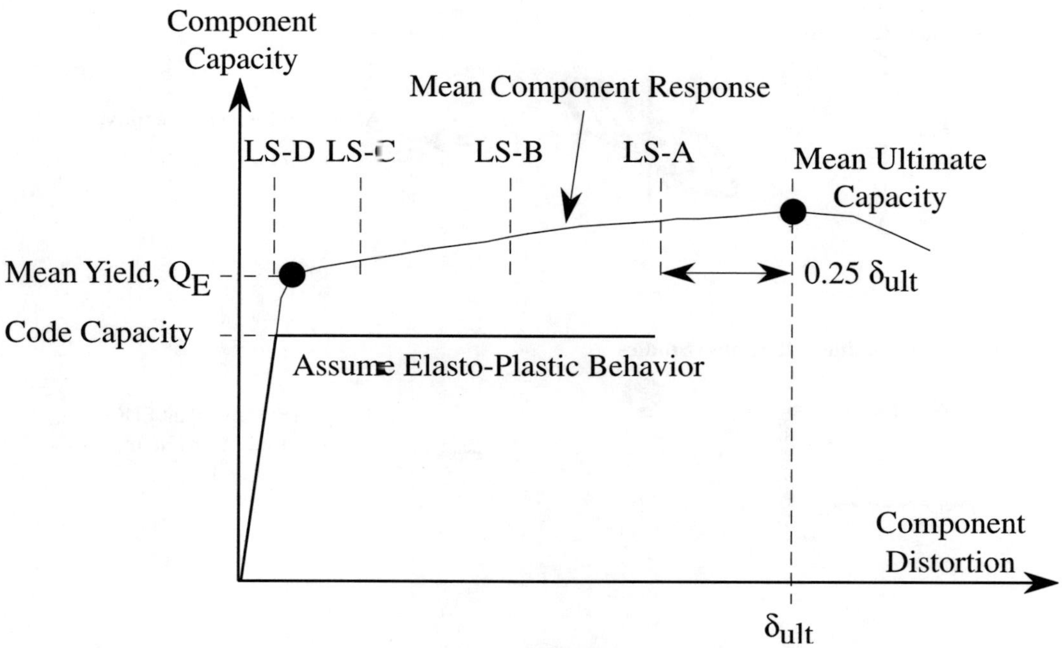

C5-4 Typical Load-Deformation Curve and Limit States

Elements that Experience Combined Response. As illustrated by Eqs. [5-1(a)] and [5-1(b)], the inelastic energy absorption factor, F_μ, is allowed only for specific response modes. Many elements have combined response modes. An example is a column subjected to combined axial force and bending. In this case, the F_μ factor is only applied to moment, generally in the following form:

$$\frac{P_{NS} + P_s}{P_{CR}} + \frac{M_{NS} + \frac{M_S}{F_\mu}}{M_P} \leq 1.0 \qquad \text{(Eq. C5-5)}$$

where

P_{NS} = Non-seismic concurrent axial force demand
M_{NS} = Non-seismic concurrent moment demand
P_S = Seismic axial force demand
M_S = Seismic moment demand
P_{CR} = Axial force capacity
M_P = Moment capacity

SECTION C5.2 ACCEPTANCE CRITERIA

This Standard utilizes force-based acceptance criteria for linear analysis and displacement-based acceptance criteria for nonlinear analysis. Both linear and nonlinear analyses are evaluated for load combinations representing the expected dead and live loads combined with the mean seismic load.

For linear analyses, the acceptance criteria is that the total demand in each element is less than or equal to the element's capacity. The seismic portion of the total element demand is reduced by an inelastic energy absorption factor, F_μ. Drift limits in Table 5-2 are specified for the total displacement. These nonlinear displacements are estimated with an elastic analysis using the load combination in Eq. (5-5).

Both story drift limits and individual element drift limits are specified for nonlinear analyses.

C5.2.3.1 Allowable Drift Limits for Structural Systems

The limits on allowable drift (Table 5-2) and on allowable rotation (Table 5-3) represent the maximum allowable deformations for a given Limit State, as shown schematically in Figure C5-4. These deformation limits were derived from Refs. [C5-6] and [C5-7] as part of the first step in F_μ development [Ref. C5-5].

Table 5-2 contains story drifts that are applied to both linear and nonlinear analyses. These drifts represent both elastic and inelastic deformations. The drift limits for braced frames and shear walls are the total deflections from the literature. The drift limits for moment frames include the nonlinear hinge rotation and an assumed elastic drift ratio of about 1/200. The drift limit for LS-D consists of the smaller of the 1/200 elastic drift ratio or the LS-C drift ratio. At these limits, cyclic strength degradation is not expected.

Table 5-3 contains nonlinear hinge rotations for rigid frame structures. The nonlinear deformations of frame elements are assumed to be lumped into nonlinear hinges, and limits on the nonlinear hinge rotation are specified. The summation of nonlinear hinge rotations and the elastic deformation of the element between hinges shall equal the total element deformation.

REFERENCES FOR SECTION C5.0

[C5-1] AISC. (1989). *Manual of steel construction: Allowable Stress Design (ASD)*, 9th Ed. Chicago.

[C5-2] AISC. (2002). "Seismic provisions for structural steel buildings." *ANSI/AISC 341-02*, Chicago.

[C5-3] AISC. (1998). *Manual of steel construction: Load and Resistance Factor Design (LRFD)*, 2nd Ed Chicago.

[C5-4] FEMA. (2000). "Recommended seismic design criteria for new steel moment frame buildings." *FEMA 350*, SAC Joint Venture for the Federal Emergency Management Agency, Washington, D.C.

[C5-5] Mertz, G.E., and Houston, T. (2001). "Force reduction factors for the structural design and evaluation of facilities containing nuclear and hazardous materials." *WSRC-TR-2001-00037*, Rev. 0. Prepared jointly by Westinghouse Savannah River Company and Structural Dynamics Engineering.

[C5-6] FEMA. (1997). "NEHRP guidelines for the seismic rehabilitation of buildings: Part 1, Provisions." *FEMA 273*, Washington, D.C.

[C5-7] Duffey, T.A., Goldman, A., Farrar, C.R. (1994). "Shear wall ultimate drift limits." *NUREG/CR-6104*, U.S. Nuclear Regulatory Commission, Washington, D.C.

[C5-8] Newmark, N.M., and Hall, W.J. (1987). *Earthquake spectra and design*. Earthquake Engineering Research Institute, El Cerrito, Calif.

[C5-9] Nassar, A.A., and Krawinkler, H. (1991). "Seismic demands for SDOF and MDOF systems." *Report No. 95*, The John A. Blume Earthquake Engineering Center, Stanford Univ.

[C5-10] Miranda, E., and Bertero, V.V. (1994). "Evaluation of strength reduction factors for earthquake-resistant design." *Earthquake Spectra*, 10(2).

[C5-11] Kennedy, R.P., et al. (1984). "Engineering characterization of ground motion: Task 1, Effects of characteristics of free-field motion on structural response." *NUREG/CR-3805*, U.S. Nuclear Regulatory Commission, Washington, D.C.

[C5-12] Osteraas, J.D., and Krawinkler, H. (1990). "Strength and ductility considerations in seismic demand." *Report No. 90*, The John A. Blume Earthquake Engineering Center, Stanford Univ.

[C5-13] Seneviratna, G. D. P. K., and Krawinkler, H. (1993). "Effects of soft soil and hysteresis model on seismic demand." *Report No. 108*, The John A. Blume Earthquake Engineering Center, Stanford Univ.

[C5-14] Seneviratna, G. D. P. K., and Krawinkler, H. (1997). "Evaluation of inelastic MDOF effects for seismic demand." *Report No. 120*, The John A. Blume Earthquake Engineering Center, Stanford Univ.

[C5-15] ATC. (1995). "Structural response modification factors." *ATC 19*, Redwood City, Calif.

[C5-16] ATC. (1995). "A critical review of current approaches to earthquake resistant design." *ATC 34*, Redwood City, Calif.

SECTION C6.0 DUCTILE DETAILING REQUIREMENTS

C6.2.2 Slab/Wall Moment Frame Systems

A slab/wall moment frame is a moment-resisting frame, composed of both walls and slabs, that resists seismic lateral loading by out-of-plane bending. Longitudinal loads are resisted by in-plane shear in the slabs and shear walls, as shown in Figure C6-1. Slab/wall moment frames constitute a very small subset of nuclear structures and are typically found in nuclear materials processing facilities where the slab and wall thickness are dictated by shielding instead of structural criteria.

Reinforced concrete structures, which resist lateral seismic load with shear walls and diaphragms in both

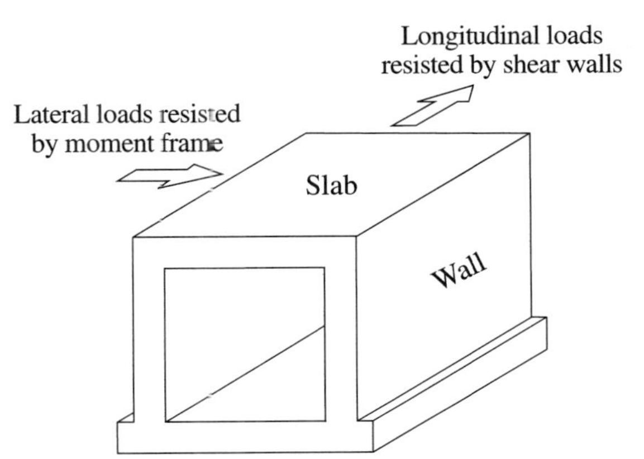

C6-1 Idealized Slab/Wall Moment Frame

directions, are not slab/wall moment frames and need not comply with the requirements of Section 6.2.2.

Below-grade walls subject to lateral soil loads with a horizontal wall span, measured between perpendicular shear walls, greater than twice the slab span, shall also be considered as slab/wall moment frames, unless it can be demonstrated that the lateral loads can be transferred to the perpendicular walls through diaphragm action.

Slab/wall moment frames typically have lower reinforcement ratios, lower shear stresses and lower axial stresses than SMRFs, which suggest that slab/wall moment frames are more ductile than SMRFs. On the other hand, slab/wall moment frames do not usually have transverse reinforcement along their entire length, and the splices may be located in hinge regions. Both of these points suggest that slab/wall moment frames are less ductile than SMRFs. Thus, slab/wall moment frames are judged to have a LS-A rotational capacity of 0.01 radians.

In slab/wall frame systems, the joints will be subjected to high shear forces, especially when the slab or the wall or both form hinges adjacent to the joint. In order to ensure structural integrity, concrete in the joint should be confined. Typical confinement details are shown in Figure C6-2.

Criterion (a) of Section 6.2.2 uses the strut-and-tie method to analyze the joint strength. One method of reinforcing a joint with the strut-and-tie method is shown in Figures C6-2(a) and C6-2(b), where diagonal crossties are used to control cracking for both opening and closing moments. The crossties should engage, at least, every other vertical rebar with a maximum bar spacing of 12 in. on center. The crossties are anchored with seismic hooks. At a minimum, the diagonal crossties should consist of #4 bars spaced 12 in. on center.

Criteria (b) and (c) of Section 6.2.2 specify slab/wall joint reinforcing which is consistent with moment-resisting frames. The horizontal ties in Figures C6-2(c) and C6-2(d) provide joint confinement. The ties should engage, at least, every other vertical rebar with a maximum bar spacing of 12 in. on center. These ties may consist of (1) hoops or (2) crossties with seismic hooks. Hoops need not overlap.

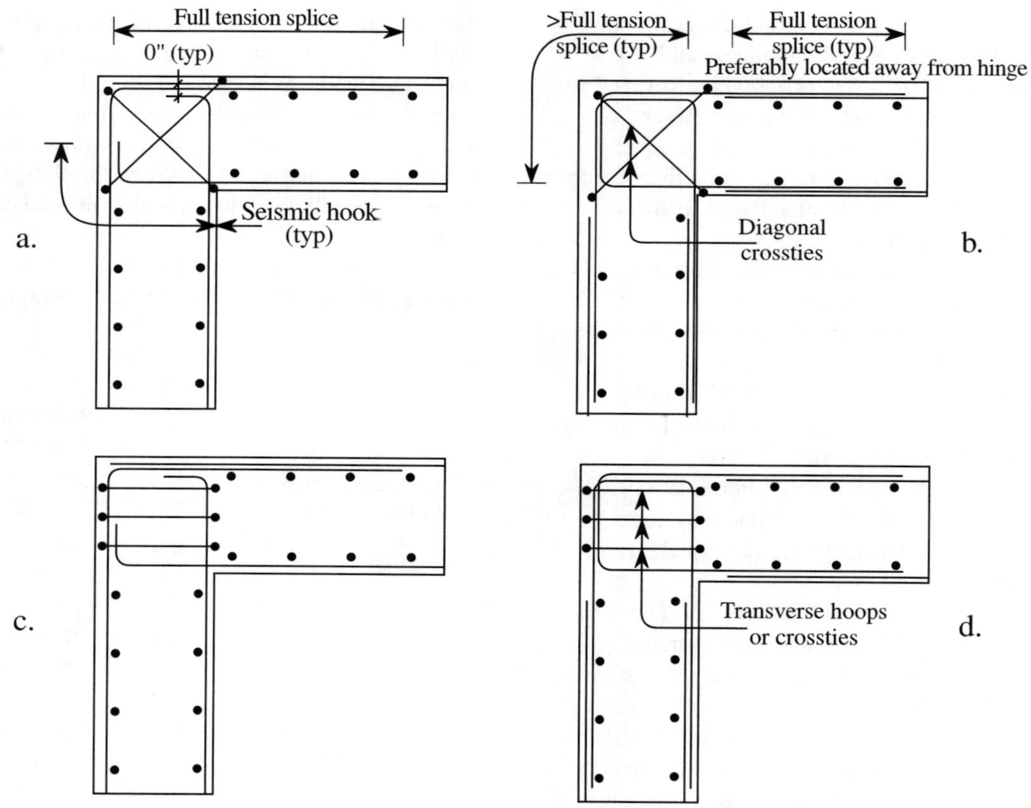

C6-2 Typical Joint Details: a and b, diagonal crossties; c and d, joint confinement

The interaction between joint shear due to frame action and joint shear due to shear-wall action (in-plane shear) may be neglected.

SECTION C7.0 SPECIAL CONSIDERATIONS

SECTION C7.1 ROCKING AND SLIDING OF UNANCHORED RIGID BODIES

Considerable uncertainty exists in the coefficient of sliding friction between an unanchored body and its sliding surface. A lower value of this coefficient of friction maximizes the sliding distance, while an upper value maximizes the rocking angle. In order to account for this uncertainty, coefficient of friction values at the estimated 95% and 5% exceedance fractiles should be estimated and used in sliding and rocking evaluations, respectively. A range of coefficients of sliding friction from 0.3 to 0.7 is considered to be reasonable for sliding of a concrete or steel rigid body on a dry, broom-finished concrete surface.

For multiple time-history inputs, each of which produces essentially the same 5% damped response spectrum, considerable variability has been found in the computed sliding distance, and moderate variability has been found in the computed rocking angle. Furthermore, if the coefficient of friction differs by as little as 0.1 between the plus and minus directions, sliding will ratchet in the direction of the lower coefficient of friction. When coupled with the variation in coefficient of friction values defined above, the factors of safety defined by Eqs. [7-1(a)] and [7-1(b)] will produce design values of sliding and rocking that have less than about a 1% probability of being exceeded for the design input motion.

SECTION C7.2 BUILDING SLIDING AND OVERTURNING

The approximate methods presented in Appendix A are strictly only applicable for sliding and rocking of rigid bodies. However, these approaches may also be used to estimate the sliding and rocking of flexible buildings, as long as the fundamental building frequency, f_b, exceeds the effective sliding and rocking frequencies, f_{eS} and f_e, respectively. When f_b does not exceed the sliding frequency, f_{eS}, a conservative estimate of the sliding displacement can be obtained by substituting f_b for f_{eS} in Eq. (A-3). When f_b does not exceed the rocking frequency, f_e, then a conservative estimate of the rocking angle, θ_o, can be obtained from Eq. [A-6(a)] by substituting f_b for f_e in this equation.

Alternatively, an acceptable approach for analyzing buildings with foundation uplift is given in Ref. [C7-1].

Typically, static resistance is the resistance about the corner, as shown in Figure 7-1. In this figure, for a rigid object on a rigid foundation, uplift occurs when the seismic load exceeds the static resistance. This does not reduce the contact area since the contact area is a point.

For a near-rigid building on soil springs, the contact area starts to reduce as soon as there is tension on one side. It continues to reduce as the seismic load increases until there is a soil or building failure due to the high contact bearing reactions.

SECTION C7.3 SEISMIC SEPARATION

Assuming that the displacements of adjacent structures are randomly phased relative to each other, the maximum probable relative displacement for the DBE ground motion can be estimated by the square root sum of squares (SRSS) of the maximum elastically computed displacements, Δ_1 and Δ_2, of the adjacent structures. The factor of 2.0 for minimum separation distance is included to provide less than a 10% chance of significant impact between adjacent structures for an input ground motion of 1.5 times the DBE. This factor was conservatively selected in lieu of requiring multiple nonlinear time-history evaluations of adjacent structures for an input ground motion of 1.5 times the DBE.

SECTION C7.5 UNREINFORCED MASONRY USED AS MOVABLE PARTITIONS, BARRIERS, AND RADIATION SHIELDING

Ref. [C7-2] provides a discussion of the lateral in-plane stiffness of masonry walls. The rapid degradation of in-plane stiffness of walls with cracking is not considered an issue, as would be the case if the walls in question were load bearing and a contributing part of the overall stiffness and in-plane load-carrying capacity of the entire structural system (Ref. [C7-3]).

Unreinforced masonry walls are used in construction of nuclear facilities where temporary or movable partitions, barriers, or radiological shielding is required. When masonry walls are intended to be permanently installed in safety-related nuclear facilities, they should be steel reinforced and anchored (Refs. [C7-3] and [C7-4]). The ACI 530-02 joint society standard (Ref. [C7-5]) is the basis for masonry wall construction for both unreinforced and reinforced masonry walls. However, when such walls are subjected to loading cases—such as design basis accident, earthquake, or wind load, which have an event probability equal to or less than 0.0005 per year—a factor of 1.6 shall be applied to the allowable stresses shown in the ACI 530-02 to achieve the Target Probability Goal in this Standard.

SECTION C7.6 PROVISIONS FOR CONSTRUCTION EFFECTS

Additional lateral wall pressures caused by soil compaction of fills adjacent to embedded walls can be estimated following the procedures described in Refs. [C7-6], [C7-7], and [C7-8]. Dynamic lateral wall pressures are typically evaluated following the elastic procedures described in Refs. [C7-9] and [C7-10].

REFERENCES FOR SECTION C7.0

[C7-1] Yim, S.C.-S., and Chopra, A.K. (1985). "Simplified earthquake analysis of multistory structures with foundation uplift." *J. Struct. Engrg.*, 111(12).

[C7-2] Sozen, M.A., et.al. (1992). "Effects of cracking and age on stiffness of reinforced concrete walls resisting in plane shear." *Paper VI/3*, North Carolina State Univ. Conf. on Nuclear Power Plant Structures, Equipment and Piping. North Carolina State University, Raleigh, N.C.

[C7-3] U.S. Dept. of Energy. (1997). "Seismic evaluation procedure for equipment in U.S. Dept. of Energy facilities." *DOE/EH-0545*, Washington, D.C.

[C7-4] ACI. (2002). "Building code requirements and commentary for masonry structures and specification for masonry structures and related commentaries." *ACI 530.1-02/530R/530.1R-02*, ACI Committee 530, Detroit.

[C7-5] ACI. (1995). "Building code requirements for masonry structures." *Joint Society Standard ACI 530-95/ASCE 5-95/TMS 402-951995*, Detroit.

[C7-6] Duncan, J. M., and Seed, R. B. (1986). "Compaction induced earth pressures under K_C conditions." *J. Geotech. Engrg.*, 112(1).

[C7-7] Terzaghi, K., Peck, R.B., and Mesri, G. (1995). Soil mechanics in engineering practice 3rd Ed. Wiley, New York.

[C7-8] Duncan, J. M., et al. (1991). "Estimation earth pressures due to compaction." *J. Geotech. Engrg.*, 117(12).

[C7-9] ASCE. (1998). "Seismic analysis of safety related nuclear structures and commentary." *ASCE 4-98*, Reston, Va.

[C7-10] Ostadan, F., and White, W.H. (1998). "Lateral seismic soil pressure: An updated approach." U.S.–Japan Soil Structure Interaction Workshop[LP6], September 22–23, U.S. Geological Survey, Menlo Park, Calif.

SECTION C8.0 EQUIPMENT AND DISTRIBUTION SYSTEMS

The philosophy used to obtain the desired performance level per the graded approach for equipment and distribution systems is described in this section.

There are four steps to qualification of equipment or distribution systems:

1. Define the level of safety and the associated safety requirements. This is addressed in ANS 2.26.
2. Develop the demand, both seismic and non-seismic
3. Determine the component capacity.
4. Compare the demand to capacity and document that the capacity is equal to or greater than the required demand.

Steps 2, 3, and 4 are addressed by this Standard.

The seismic motion at the support of the equipment and distribution systems (i.e., components) is based on an elastic analysis of the supporting structure for the DBE that is defined by the SDC of the component. This elastic seismic analysis is performed per Section 3.0 of this Standard, which refers to seismic analysis procedures of ASCE 4. ASME B&PVC, Section III, Nonmandatory Appendix N, can also be used. The inelastic energy absorption factor, F_{μ}, of the supporting structure is set to 1.0 unless it can be justified, considering the potential structure over strength, that the inelastic response of the supporting structure will effectively lower the support motion. The F_{μ} values in Table 8-1 are consistent with R_p values referenced in Chapter 16 of IBC 2003 (Table 9.6.2.2 of ASCE 7-02) when it is understood that the allowable stresses for ASME Service Level D may be as high as twice yield, while the IBC and other structure code allowables are typically no more than yield. The guidance of ASCE 4-98, Section 3.4, "Input for Subsystem Seismic Analysis," can be used to define the input motion including reduction of spectral amplitudes.

The application of equipment and distribution systems inelastic energy absorption factor, F_{μ}, provided in Table 8-1, is defined by the specified Limit State and is only to be used for qualification by analysis. F_{μ} is not used when qualification is performed by test or experience data methods since the Limit State is indirectly specified for the component by physical observed behavior (e.g., functional requirements, structural integrity) that defines the acceptance criteria.

The capacity of the component qualified by analysis is based on the specified code provisions and the allowable stress limits, such as the ASME Service Level limits. Specified codes and standards are listed in Table C8-1 as a function of type of component and the specified SDC. Within many of these standards, different ac-

TABLE C8-1. Typical Classification and Standards Used for Construction and Procurement of Mechanical and Electrical Equipment

Items	NRC RG 1.26 CLA ANS 58.14 QC-1 SDC-5–ANS 2.26	NRC RG 1.26 CLB ANS 58.14 QC-2 SDC-4–ANS 2.26	NRC RG 1.26 CLC ANS 58.14 QC SDC-3–ANS 2.26
A. MECHANICAL ITEMS			
1. Distribution systems			
Piping (process–critical)	ASME B&PVC, Sec. III, Cl. 1	ASME B&PVC, Sec. III, Cl. 2	ASME B&PVC, Sec. III, Cl. 3
Piping (process–other)	ASME B&PVC, Sec. III, Cl. 2	ASME B&PVC, Sec. III, Cl. 3	ASME B&PVC, Sec. III, Cl. 3
Duct (HVAC)–round	ASME B&PVC, Sec. III, Cl. 2, or AG-1, SA	ASME B&PVC, Sec. III, Cl. 3, or AG-1, SA	SMACNA
Duct (HVAC)–rectangular	AG-1, SA	AG-1, SA	SMACNA
Duct (process)–round	ASME B&PVC, Sec. III, Cl. 2, or AG-1, SA	ASME B&PVC, Sec. III, Cl. 3, or AG-1, SA	SMACNA
Duct (process)–rectangular	AG-1, SA	AG-1, SA	SMACNA
Tubing	ASME B&PVC, Sec. III, Cl. 3	ASME B31.1 or B31.3	Mfr. spec.
2. Passive components			
Pressure vessels (critical)	ASME B&PVC, Sec. III, Cl. 1	ASME B&PVC, Sec. III, Cl 2	ASME B&PVC, Sec. III, Cl. 3
Pressure vessels (other)	ASME B&PVC, Sec. III, Cl. 2	ASME B&PVC, Sec. III, Cl 3	ASME B&PVC, Sec. VIII, Div. 2
Heat exchangers (vital)	ASME B&PVC, Sec. III, Cl. 1	ASME B&PVC, Sec. III, Cl 2	ASME B&PVC, Sec. III, Cl. 3
Heat exchangers (other)	ASME B&PVC, Sec. III: Cl. 1–Primary Cl. 2–Secondary	ASME B&PVC, Sec. III: Cl. 2–Primary Cl. 3–Secondary	ASME B&PVC, Sec. VIII, Div. 2
Tanks (water, atmospheric–critical)	ASME B&PVC, Sec. III, ND-3900	ASME B&PVC, Sec. III, ND-3900	API 650
Tanks (water, atmospheric–other)	ASME B&PVC, Sec. III, ND-3900	API 650	AWWA 100
Tanks–press. 0–15 psig (vital)	ASME B&PVC Sec. NC-3800	ASME B&PVC, Sec. III, ND-3800	API 620
Tanks–press. 0–15 psig (other)	ASME B&PVC, Sec. ND-3800	API 620	API 620
Glove boxes	ANSI/ASTM C852, ANS 11.16, AISC N690	ANSI/ASTM C852, ANS 11.16, AISC N690	ANSI/ASTM C852, ANS 11.16, AISC N690
3. Active components			
Pumps (critical)	ASME B&PVC, Sec. III, Cl. 1	ASME B&PVC, Sec. III, Cl. 2	API 610
Pumps (other)	ASME B&PVC, Sec. III, Cl. 2	ASME B&PVC, Sec. III, Cl. 3	API 610
Circulators	ASME AG-1, BA	ASME AG-1, BA	SMACNA
Fans	ASME AG-1, BA	ASME AG-1, BA	SMACNA
Valves–(critical)	ASME B&PVC, Sec. III, Cl. 1	ASME B&PVC, Sec. III, Cl. 2	ASME B&PVC, Sec. III, Cl. 3
Valves–(other)	ASME B&PVC, Sec. III, Cl. 2	ASME B31.1	ASME B&PVC, Sec. III, Cl. 3
Dampers	ASME AG-1, BA	ASME AG-1, BA	SMACNA
Cranes and hoists (critical)	ASME NOG-1 or NUM-1, Type 1	ASME NOG-1 or NUM-1, Type II	CMAA 70
Cranes and hoists (other)	ASME NOG-1 or NUM-1, Type II	ASME NOG-1 or NUM-1, Type II	CMAA 70

(continues)

TABLE C8-1 Typical Classification and Standards Used for Construction and Procurement Of Mechanical and Electrical Equipment *(Continued)*

Items	NRC RG 1.26 CLA ANS 58.14 QC-1 SDC-5–ANS 2.26	NRC RG 1.26 CLB ANS 58.14 QC-2 SDC-4–ANS 2.26	NRC RG 1.26 CLC ANS 58.14 QC SDC-3–ANS 2.26
Engines	IEEE 387	IEEE 387	Mfr. spec.
Compressors	ASME B&PVC, Sec. III, Cl. 3	ASME B&PVC, Sec. III, Cl. 3	Mfr. spec.
B. ELECTRICAL ITEMS			
1. Distribution systems			
Cable trays	IEEE 628 NEMA VE-1	IEEE 628 NEMA VE-1	Mfr. spec.
Conduit	ASME B&PVC, Sec. III, Cl. 3	ASME B31.1 or B31.3	Mfr. spec.
2. Passive components			
Transformers	IEEE 638	IEEE 638	IEEE C-57.12 Series
Batteries	IEEE 450 IEEE 323 IEEE 344 IEEE 535	IEEE 535 IEEE 450	Mfr. spec.
Battery racks	IEEE 450 AISC	IEEE 450 AISC	IEEE 450 AISC
Electrical panels and racks	IEEE 420 NEMA IC-6	IEEE 420 NEMA IC-6	Mfr. spec.
3. Active components			
Motor control centers	IEEE 649	IEEE 649	IEEE 649
Switch gear	IEEE C-37.82	IEEE C-37.82	Mfr. spec.
Motors	IEEE 334 IEEE 308 IEEE 323 IEEE 344	IEEE 334 IEEE 308	Mfr. spec.
Valve actuators	IEEE 382	IEEE 382	Mfr. spec.
Generators	IEEE 387 IEEE 308 IEEE 323 IEEE 344	IEEE 387 IEEE 308	Mfr. spec.
Battery chargers and inverters	IEEE 382	IEEE 382	Mfr. spec.
E. INSTRUMENTATION AND CONTROL (g)			
Relays	IEEE 37.98	IEEE 37.98	IEEE 37.98
Switches	IEEE 344	IEEE 344	IEEE 344
Pressure sensors	IEEE 344	IEEE 344	IEEE 344
Temperature sensors	IEEE 344	IEEE 344	IEEE 344
Radiation sensors	IEEE 344	IEEE 344	IEEE 344
Radiation alarms	ANSI 13.1 ANSI 42.18	ANSI 13.1 ANSI 42.18	ANSI 13.1 ANSI 42.18
F. EQUIPMENT SUPPORTS & ANCHORS			
Equipment support–steel	AISC, ASME NF, AISI	AISC, ASME NF, AISI	AISC, AISI
Equipment support–concrete	ACI 349	ACI 349	ACI 349
Equipment anchor–concrete	ACI 349, App. B	ACI 349, App. B	ACI 349, App. B
Dist. syst. std. supports	MSS SP-58	MSS SP-58	MSS SP-58

Notes:

Critical SSCs are those that normally contain or distribute significant amounts of radioactive or other toxic substances. Such items are typically limited to systems with design temperatures greater than 300°F and design pressures greater than 500 psi, piping with diameter larger than 4.0 in., and vessels larger than 150 ft^3. It also applies to systems and components within a given safety-related class or performance category that are considered particularly important to safety.

See Sections 4.2.2 and 4.2.4 for appropriate structural capacities in accordance with ACI 349 and AISC.

ceptance criteria are presented as a function of the specified Limit State. For passive components, the stress limits are based so they are not more conservative than that required for safety-related equipment in nuclear power plants (NPP) when designing for the NPP's SSE. This is essentially SDC-5–Limit State D. For mechanical components, this is typically ASME B&PVC, Section III, Service Level D. Therefore, Section 8.0 allows Service Level D for passive components for all Limit States. Of course, the seismic input will vary depending on the component's SDC. For active components (must move or otherwise change state), the allowable stress level (e.g., ASME B&PVC, Section III, Service Level) must be more conservative to ensure the active functional requirements are met. This can be ensured by maintaining all stress levels less than yield and meeting the displacement criteria. For active components, Service Level A or B of ASME B&PVC, Section III, is often specified. For Service Level B allowables, primary stresses are typically limited to 0.8 to 1.0 times the specified minimum yield at temperature, with secondary stresses limited to twice yield. F_μ is set equal to 1.0 for calculating displacements in order to obtain true estimates of these displacements.

Table C8-1 lists many acceptable codes and standards. Some present alternative methods for seismic qualification of a component. For example, IEEE 628 (Ref. [C8-12]) provides guidance on acceptable qualification methods for electrical distribution systems, including analysis, dynamic testing, and seismic experience-based qualification.

Component seismic qualification methods—specified in IEEE 344 (Ref. [C8-4]), ASME QME-1 (Ref. [C8-5]), and others, as listed in Table C8-1—are acceptable within the specified limitations provided in Section 8.0. When using test or experience data, such as TES and earthquake experience, an equipment capacity factor has to be considered in order to obtain an equal confidence level of performance. The equipment capacity factor of 1.4 for test and TES and 1.0 for earthquake experience are based on Ref. [C8-10]. The seismic Scale Factors, *SF*, presented in this study have been explicitly included in the specified ground motion when using the required Design Factor, *DF*, given in Section 2.2.1.

Experience-based approaches for verifying seismic adequacy of certain components have been successfully implemented for more than 10 yr. The Seismic Qualification Utility Group (SQUG) Generic Implementation Procedure (GIP) (Refs. [C8-6] to [C8-9]) and the DOE Seismic Evaluation Procedure (Ref. [C8-11]) are examples of these approaches. But those who apply these experience-based approaches are required to meet certain experience requirements and have formal training in the application of these methods. At the time of the writing of this Standard, revisions to IEEE 344 and ASME QME-1 to incorporate experience-based approaches are being balloted and may be used when approved for experience-based methods of seismic qualification.

The effect of seismic interaction of adjacent SSCs would be part of the final assurance of seismic qualification of the safety-related item. Interaction of non-safety SSCs that could affect the safety function of equipment or distribution systems would typically have the same SDC but probably a lower Limit State.

REFERENCES FOR SECTION C8.0

The following lists references associated with Table C8-1 and other documents presented in Section 8.0 and this Commentary.

[C8-1] International Code Council, Inc. (2003). *International Building Code 2003*, Chap. 16. Whittier, Calif.

[C8-2] ASME. (2004). *Boiler and Pressure Vessel Code (B&PVC)*. Section III, Appendix N, "Dynamic analysis methods." New York.

[C8-3] ASCE. (2000). "Seismic analysis of safety-related nuclear structures and commentary." *ASCE 4-98*, Reston, Va.

[C8-4] IEEE. (1987). "Recommended practices for seismic qualification of class 1E equipment for nuclear power generating stations." *IEEE 344*, New York.

[C8-5] ASME. (2000). "Qualification of active mechanical equipment used in nuclear power plants." *ASME QME-1*, New York.

[C8-6] Electric Power Research Institute (EPRI). (1991). "Summary of the seismic adequacy of twenty classes of equipment required for the safe shutdown of nuclear plants." *EPRI NP-7149-D*, Palo Alto, Calif.

[C8-7] Seismic Qualification Utility Group (SQUG). (1997). "Generic Implementation Procedure (GIP) for seismic verification of nuclear plant equipment," Rev. 3, Electric Power Research Institute, Palo Alto, Calif.

[C8-8] U.S. Nuclear Regulatory Commission. (1992). "NRC Supplemental Safety Evaluation Reports on the GIP." *SSER No. 2*, Rev. 2, corrected 2/14/92; *SSER No. 3*, Rev. 3, updated 5/16/97.

[C8-9] Seismic Qualification Utility Group (SQUG). (2000). "Implementation guidelines for seismic qualification of New and Replacement Equipment/Parts (NARE) using the Generic Implementation Procedure (GIP)," Rev. 4, Electric Power Research Institute, Palo Alto, Calif.

[C8-10] Salmon, M.W., and Kennedy, R.P. (1994). "Meeting performance goals by use of experience data." *UCRL-CR-120813*. Prepared for Lawrence Livermore National Laboratory and U.S. Dept. of Energy Existing Facilities Project Steering Group and Technical Review Team, Livermore, Calif.

[C8-11] U.S. Dept. of Energy. (1997). "Seismic evaluation procedure for equipment in U.S. Dept. of Energy facilities." *DOE/EH-0545*, Washington, D.C.

[C8-12] IEEE. (2001). "Standard criteria for the design, installation, and qualification of raceway systems for class 1E raceway systems for class 1E circuits for nuclear power generating stations." *IEEE 628*, New York.

[C8-13] Manufacturers Standardization Society of the Valve and Fitting Industry. (2002). "Pipe hangers and supports: Materials, design and manufacture." *MSS SP-58,* Vienna, Va.

[C8-14] ASME. (2004). *Boiler and Pressure Vessel Code (B&PVC)*. Section III, Subsection NF, "Supports." New York.

[C8-15] ACI. (2002). "Building code requirements for reinforced concrete." *ACI 318*, Detroit.

[C8-16] ACI. (2001). "Code requirements for nuclear safety related concrete structures." *ACI 349*, Detroit.

[C8-17] AISC. (1989). "Specification for the design, fabrication and erection of structural steel for buildings." *AISC Allowable Stress Design Manual*, Chicago.

[C8-18] AISC. (1994). "Specification for the design, fabrication, and erection of steel safety related structures for nuclear facilities." *AISC N690*, Chicago.

[C8-19] AISI. (1996). "Specification for design of cold formed steel structural members." SG-2000-1, Washington, D.C.

[C8-20] ANS. (1988). "Design guide for radioactive material handling facilities and equipment." *ANS 11.16*, La Grange Park, Ill.

[C8-21] API. (2003). "Centrifugal pump for general refinery service." *API 610*, Washington, D.C.

[C8-22] API. (2002). "Design and construction of large, welded, low-pressure storage tanks." *API 620*, Washington, D.C.

[C8-23] API. (1998). "Atmospheric welded steel tanks for oil storage." *API 650*, Washington, D.C.

[C8-24] ASME. (2002). "Ventilation air cleaning and air conditioning fans and blowers." *ASME AG-1, Div. II, BA*, New York.

[C8-25] ASME. (2002). "Ventilation air cleaning and air conditioning dampers and louvers." *ASME AG-1, Div. II, SA*, New York.

[C8-26] ASME. (2004). "Power piping." *ASME B31.1*, New York.

[C8-27] ASME. (2004. *Boiler and Pressure Vessel Code (B&PVC)*. Section III, Cl. 1, "Nuclear components subsection NB." New York.

[C8-28] ASME. (2004). *Boiler and Pressure Vessel Code (B&PVC)*. Section III, Cl. 2, "Nuclear components subsection NC." New York.

[C8-29] ASME. (2004). *Boiler and Pressure Vessel Code (B&PVC)*. Section III, Cl. 3, "Nuclear components subsection ND." New York.

[C8-30] ASME. (2004). *Boiler and Pressure Vessel Code (B&PVC)*. Section VIII, Div. 1, "Unfired pressure vessels." New York.

[C8-31] ASME. (2004). *Boiler and Pressure Vessel Code (B&PVC)*. Section VIII, Div. 2, "Unfired pressure vessels design by analysis." New York.

[C8-32] ASME. (2002). "Standards on nuclear type I overhead, underhead, gantry and monorail cranes." ASME *NOG-1, Type I*, New York.

[C8-33] ASME. (2002). "Standards on nuclear type II overhead, underhead, gantry and monorail cranes." *ASME NOG-1, Type II*, New York.

[C8-34] ASTM. (1997). "Standard guide for design criteria for plutonium glove boxes." *ASTM C 852-93 (R1997)*, West Conshohocken, Pa.

[C8-35] American Water Works Association (AWWA). (1996). "Welded steel tanks for water storage." *AWWA D100-96*, Denver.

[C8-36] Crane Manufacturer Association of America (CMAA). (2004). "Specification for electric overhead traveling cranes." *CMAA 70*, Charlotte, N.C.

[C8-37] IEEE. (2001). "Standard criteria for class 1E power systems for nuclear power generating stations." *IEEE 308*, New York.

[C8-38] IEEE. (2003). "Standard for qualifying class 1E equipment for nuclear power generating stations." *IEEE 323*, New York.

[C8-39] IEEE. (1994). "Type test of continuous duty class 1E equipment for nuclear power generating stations." *IEEE 334*, New York.

[C8-40] IEEE. (1987). "Recommended practice for seismic qualification of class 1 equipment." *IEEE 344*, New York.

[C8-41] IEEE. (1996). "Standard for qualification of actuators for power operated valve assemblies with safety related functions for nuclear power plants." *IEEE 382*, New York.

[C8-42] IEEE. (1995). "Criteria for diesel-generator units applied as standby power supplies for nuclear power generating structures." *IEEE 387*, New York.

[C8-43] IEEE. (2001). "Standard design and qualification of class 1E control boards, panels and racks used in nuclear power generating stations." *IEEE 420*, New York.

[C8-44] IEEE. (2002). "Recommended practice maintenance, testing and replacement of large lead storage batteries." *IEEE 450*, New York.

[C8-45] IEEE. (1986). "Standard qualifications of class 1E lead storage batteries for nuclear power generating stations." *IEEE 535*, New York.

[C8-46] IEEE. (1992). "Standard for qualifications of class 1E transformers for nuclear generating stations." *IEEE 638*, New York.

[C8-47] IEEE. (1991). "Standard for qualifying class 1E motor control centers for nuclear power generating stations." *IEEE 649*, New York.

[C8-48] IEEE. (1987). "Standard for the qualification of switchgear assemblies for class 1E applications in nuclear power generating stations." *IEEE C-37.82*, New York.

[C8-49] IEEE. (1987). "Standard for seismic testing of relays." *IEEE C-37.98*, New York.

[C8-50] IEEE. (2002). "Standards for transformers." *IEEE C-57.12 Series*, New York.

[C8-51] Manufacturers Standardization Society of the Valve and Fitting Industry. (2002). "Pipe hangers and supports: Materials, design and manufacture." *MSS SP-58,* Vienna, Va.

[C8-52] National Electrical Manufacturer Association (NEMA). (1993). "Industrial control and systems: enclosures." *NEMA ICS-6*, Rosslyn, Va.

[C8-53] National Electric Manufacturer Association (NEMA). (2002). "Metallic cable tray systems." *NEMA VE-1*, Rosslyn, Va.

[C8-54] Sheet Metal and Air-Conditioning Contractors National Association (SMACNA). (1990). "HVAC systems duct design." Chantilly, Va.

[C8-55] Sheet Metal and Air-Conditioning Contractors National Association (SMACNA). (2004). "Rectangular industrial duct construction standard." Chantilly, Va.

[C8-56] Sheet Metal and Air-Conditioning Contractors National Association (SMACNA). (2002). "Round industrial duct construction standard." Chantilly, Va.

[C8-57] ANS. (1993). "Safety and pressure integrity classification criteria for light water reactors." *ANSI/ANS-58.14*, La Grange Park, Ill.

[C8-58] US Nuclear Regulatory Commission. (1976). "Quality group classifications and standards for water-, steam-, and radioactive-waste-containing components of nuclear power plants." *RG1.26.*

SECTION C9.0 SEISMIC QUALITY PROVISIONS

SECTION C9.1 DESIGN VERIFICATION AND INDEPENDENT PEER REVIEW

As a minimum, the independent peer review shall include the geophysical and geotechnical design basis and the seismic analysis and design of the SSCs.

SECTION C9.2 STRUCTURAL OBSERVATION, INSPECTION, AND TESTING

The graded approach for construction quality should be developed based on the SDC assigned. The minimum for SDC-3 should be IBC 2003 and range to an equivalent of commercial nuclear requirements as outlined in ASME NQA-1, Part 2. Additional guidance on developing inspection requirements can be obtained from ASCE Manual No. 73. As a minimum, the testing and inspection requirements of the applicable material code (i.e., ACI 349, N690, etc.) shall be specified.

SECTION C9.3 QUALITY ASSURANCE

In developing the details of the QA program to meet the DOE or NRC requirements, refer to their guidance documents (e.g., DOE G 414.1-2, Quality assurance management system guide). Nuclear facility design and seismic analysis detailed guidance is also available in DOE and NRC guidance documents. Chapter 17 of Ref. [C9.3] provides useful input for commercial facilities and should aid in developing a graded approach.

The DOE QA requirements for nuclear facilities are defined in 10 CFR 830, Subpart A, Quality Assurance. DOE also adopts nonmandatory design process guidelines specifically for seismic analysis in supporting standards. The NRC QA requirements for nuclear power facilities are established in 10 CFR Part 50, Appendix B.

REFERENCES FOR SECTION C9.0

[C9.1] ASME. (2002). *Quality Assurance Requirements for Nuclear Facility Applications, NQA-1*, New York.

[C9.2] ASCE. (2000). "Quality in the Constructed Project: A Guide for Owners, Designers, and Constructors." ASCE Manuals and Reports on Engineering Practice, No. 73, Reston, Va.

[C9.3] International Code Council, Inc. (2003). *International Building Code 2003*, Whittier, Calif.

[C9.4] DOE G 414.1-2, Quality assurance management system guide, U.S. Department of Energy, Washington, D.C.

[C9.5] 10 CFR 830, Subpart A, Quality Assurance

[C9.6] 10 CFR Part 50, Appendix B

INDEX